SAVE HARMLESS AGREEMENT

Because use of the information, instructions and materials discussed and shown in this book, document, electronic publication or other form of media is beyond our control, the purchaser or user agrees, without reservation to save Knowledge Publications Corporation, its agents, distributors, resellers, consultants, promoters, advisers or employees harmless from any and all claims, demands, actions, debts, liabilities, judgments, costs and attorney's fees arising out of, claimed on account of, or in any manner predicated upon loss of or damage to property, and injuries to or the death of any and all persons whatsoever, occurring in connection with or in any way incidental to or arising out of the purchase, sale, viewing, transfer of information and/or use of any and all property or information or in any manner caused or contributed to by the purchaser or the user or the viewer, their agents, servants, pets or employees, while in, upon, or about the sale or viewing or transfer of knowledge or use site on which the property is sold, offered for sale, or where the property or materials are used or viewed, discussed, communicated, or while going to or departing from such areas.

Laboratory work, scientific experiment, working with hydrogen, high temperatures, combustion gases as well as general chemistry with acids, bases and reactions and/or pressure vessels can be EXTREMELY DANGEROUS to use and possess or to be in the general vicinity of. To experiment with such methods and materials should be done ONLY by qualified and knowledgeable persons well versed and equipped to fabricate, handle, use and or store such materials. Inexperienced persons should first enlist the help of an experienced chemist, scientist or engineer before any activity thereof with such chemicals, methods and knowledge discussed in this media and other material distributed by KnowledgePublications Corporation or its agents. Be sure you know the laws, regulations and codes, local, county, state and federal, regarding the experimentation, construction and or use and storage of any equipment and or chemicals BEFORE you start. Safety must be practiced at all times. Users accept full responsibility and any and all liabilities associated in any way with the purchase and or use and viewing and communications of knowledge, information, methods and materials in this media.

BY D. S. HALACY, JR.

FABULOUS FIREBALL

The Story of Solar Energy

Illustrated with photographs

THE MACMILLAN COMPANY
New York

For My Daughter, Jessie Ann

ISBN:978-1-60322-029-3

The Macmillan Company, New York
Collier-Macmillan Canada, Ltd., Toronto, Ontario
Library of Congress catalog card number: 57-10014
Printed in the United States of America

Fourth Printing 1966

Contents

INTRODUCTION

One hundred years ago, when steam engines and water wheels were the only means available for producing power, men in the United States and in many other nations sought ways to use the sun's energy. Less than fifty years ago, great solar engines were built in Arizona and California to pump water for irrigation. The largest of all solar machines was finished in Egypt just before World War I, to raise water from the River Nile.

When the steam turbine and the gasoline engine became available virtually everywhere, these solar engines were discarded and their inventors were all but forgotten. Only in those areas of the world where fuels were scarce did men continue to dream of ways to use the unending flow of energy from the sun. Meanwhile, the demand for conventional fuels—wood, coal, oil, and natural gas—increased rapidly. Today, more and more energy is needed for automobiles and airplanes, farms and factories, and to heat and cool our homes.

Once again, men are beginning to give serious thought to ways and means for using the sun's energy. We know that we are consuming our supply of fossil fuels with alarming rapidity, and even our newest energy source, atomic fission, depends upon a relatively limited supply of uranium and thorium. Solar radiation remains as the world's only inexhaustible source of energy, but before we reach the point where we can make use of this endless resource, many discoveries will have to be made. Energy storage methods, large area solar-electric cells, catalysts for photo-activated decomposition of water—these are some of the fields where new knowledge is needed.

These discoveries will in all probability be made by men and women who are today still boys and girls in school. The scientists of tomorrow who turn to the study of the sun will do so because their interest and curiosity have been aroused by books such as Mr. Halacy's FABULOUS FIREBALL. He gives a good background on the scientific aspects of solar energy, and his glimpse of what may come in the future is intriguing. The Association for Applied Solar Energy welcomes this addition to the growing body of information on the sun and the ways in which mankind will make use of it.

John I. Yellott,
Executive Director
Association for Applied Solar Energy

May, 1957

CHAPTER
ONE

Meet the Sun

In a very real sense, we were born of the sun. Ages ago our earth was wrenched from its parent and hurled a tremendous distance away. Today, we still look to the sun for nurture; we live in the sun or die, as one writer has said. The sun has given everything, including a name, to our solar system.

The sun holds us in an orbit, keeping us from flying off into outer space and disaster. It warms us, and grows our crops. We see by its light, and it causes our weather. Sailing ships, windmills, and waterfalls all owe their existence to the sun. Sunbathers and navigators alike depend on the sun, though for very different reasons.

Quite recently in our history, men worshiped the sun, and an eclipse was enough to terrify superstitious natives who feared that such a thing foretold the end of the world. Today, we no longer worship the sun but serious-minded scientists eye it with a new and growing respect.

When you turn on television, cook dinner, or drive the family car across town, stop and think where the power comes from. Electricity, natural gas, and petroleum of course; but all these fuels come from something much more basic. The source of power is that ball of fire up in the sky some ninety-three million miles away. Quite literally the sun makes the world go 'round!

Every ounce of energy in wood, coal, oil and gas was put there over a long period of time by the sun shining on the earth and storing up heat. In fact, we refer to these as "fossil fuels" to indicate how long it took to create them. When we burn fossil fuels we simply release stored energy, much like running water from a tank.

Once men thought these fuels would last forever. Forests were stripped and no one was too concerned, for coal and petroleum deposits were being found faster than they could be used. Now, however, the picture has changed completely. Experts believe that our known reserves of all types of fuel, including the atomic materials, will last only a few hundred years *at present rates of consumption.*

These scientists know too that man is using power in greatly increasing amounts each year, so present consumption is not a realistic method of computing the time we have left. For example, jet airplanes gulp fuel much faster than the old types did—one new fighter plane has a horsepower rating of 60,000! And guided missiles and rockets consume fuel in a way to make conservationists fearful.

Our automobiles now use more fuel as engines grow larger; even closer to home are the other demands we make. Comfortably heated and cooled houses have become necessities today, and power-consuming appliances continue to grow in popularity. All over the world, but

Association For Applied Solar Energy

SCOUTS LEARNING basic facts about solar energy. Much of it is reflected or absorbed by the atmosphere. Rain, and the rest of our weather, depends on the sun.

particularly in America, people have become "power-mad" in a different meaning of the term.

Modern technology is taxing our dwindling fuel supplies. As it becomes increasingly necessary to use low-grade ores, for example, more power is required to process metal. Our chemical industry cuts into petroleum and coal reserves, with detergents and synthetic fibres taking the place of "natural" materials. Then too, work animals are being replaced with power equipment. This further depletes the fossil fuel reserve.

There is another big reason why the bottom of the fuel barrel is nearly in sight. Population is increasing tremendously and is expected to *triple* in the next century. What this will do to the stockpile is as alarming as it is unavoidable. A hundred and fifty years ago an Englishman named Malthus warned that existing methods could not feed and support much increase in population. Improved methods and industrialization met the challenge then, but we are now faced with a grimmer prospect.

In Malthus' day, only the surface of the world's fuel supply had been scratched. Today prospectors are scrambling to find new supplies. Even if the optimists are right and huge new supplies of fossil fuels are discovered along with more atomic materials, the result will be only a delaying of the inevitable result. This is because of one inescapable fact: *Fossil fuels and atomic materials, once used, are gone forever!* Man cannot wait the ages necessary for nature to again build him an easy-to-use stockpile of what scientists call "chocolate-bar" fuels.

The use of simple woodburning engines has been suggested as a possible power supply. Trees grow rapidly, and fairly efficient engines might be made. However it is obvious that enough wood cannot be grown to supply the whole world even if the transportation problem could be solved. Taking all factors into consideration, responsible men predict that we will begin to feel the pinch of power shortage as early as 1975.

This is the prime reason for interest in solar energy, for if man cannot wait for stored, secondhand power, he must use it directly from the sun. For an idea of the tremendous possibilities, consider this: *Every three days, the sun showers the earth with the equivalent energy of all the stored fuel reserves known!* To better understand this amazing fact, let's take a good look at Sol.

Our sun is actually an orange dwarf "star." Although it is not nearly the hottest star known, its surface has a temperature of about 10,000 degrees F. and its interior is thought to be in the range of twenty *million* degrees. We commonly think of the sun as burning, yet it is too hot to burn and is composed of elemental gases. On the surface this gas is thin and tenuous, but so great is the gravity of the sun that at its center it has a density six times that of the element, mercury.

The sun is 864,000 miles in diameter, and has a mass one third of a million times greater than earth. Scientists believe it is two billion years old, and that instead of cooling, it is still getting hotter. Perhaps in another two billion years it will have reached a temperature sufficient to destroy the earth. Long after that, the sun will wane and gradually cool.

For centuries men have pondered the source of the fantastic quantities of heat and light the sun beams into space. Now it is thought that the fusion of hydrogen atoms results in the formation of helium, with heat energy as a by-product. Helium, the gas used in dirigibles and balloons, was first discovered in the sun and takes its name from the Greek word for sun.

The sun's corona, or halo, is almost as hot as its interior. Solar prominences—tongues of hot gas—leap outward from the surface a half-million miles at speeds reaching 250,000 miles an hour. Fortunately these prominences do not reach the ninety-three million miles to earth, nor are the terrific temperatures transmitted through space. Some of the energy from these disturbances does reach our atmosphere however, and is believed to cause changes in weather.

Using an instrument called a coronagraph, scientists observe the east "limb" of the sun. Since the sun rotates once in twenty-eight days, the sector observed moves to face the earth in a week and makes it possible to predict general weather changes that far ahead.

Besides weather, solar activity affects radio reception too. The Aurora Borealis, or northern lights, are caused by the glowing of gases high in our atmosphere when they are struck by energy particles from the sun. These things are provable, and Dr. C. G. Abbot, dean of solar scientists, feels that the sun's influence is much more basic than is generally realized. He compares its cycles to those of human life. Other researchers claim to show a correlation between wars in past centuries and sunspot activity! Perhaps in time these things will be proved; for the present it is interesting enough to know that the sun is constantly bombarding us with "bullets" of heat energy that we can put to work.

The saying that there is nothing new under the sun has

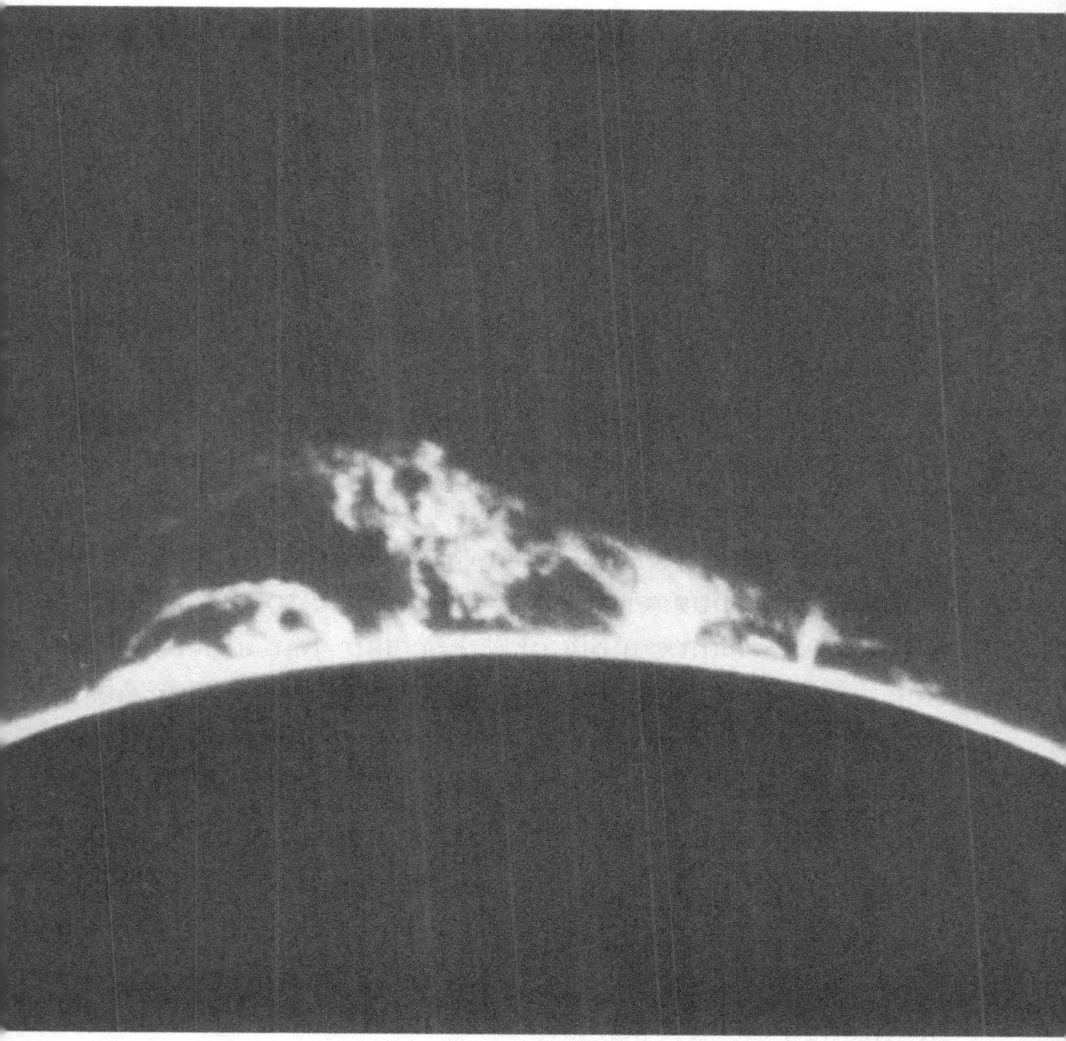

High Altitude Observatory, University of Colorado

SOLAR PROMINENCES leaping from the surface of the sun. These tongues of hot gas extend thousands of miles outward. This photograph was taken with the coronagraph of the High Altitude Observatory of the University of Colorado.

some truth in it. A hundred years ago men built steam engines using solar heat, and Samuel Pierpont Langley, whose flying machine vied with the Wrights', built a solar oven which he took to the top of Mt. Whitney to cook a meal.

Despite these and other experiments, interest in solar energy lagged. Other forms of power were simpler to use, so man used them. Wood was burned, then coal, oil and gas. Finally electricity was made from other fuels, and with seemingly endless reserves there was no pressure to do otherwise.

Now it appears that our wastefulness has caught up with us. The hunger for power and the sobering knowledge of the scant supply of "chocolate bar" fuel is spurring solar research. Fortunately we can be optimistic. Although the energy reaching the earth is only one thousand millionth of the sun's total, it is still two thousand times the requirements of even those of us in America. We are the biggest power-users in the world, and yet it has been estimated that the roof of the average house receives and wastes many times more energy than comes in through electric wires!

One acre of ground in the southern United States gets 4,000 horsepower during the hours of sunlight, and the solar energy falling on Lake Mead is five times greater than power generated in the Boulder Dam hydroelectric plant. This sun energy is "diffuse," or at a low concentration, and will challenge the ingenuity of the "heliotechnologist" to use it efficiently. But use it we must, or someday do without.

A different kind of hunger in some parts of the world gives impetus to solar research. In lands where food and

Association For Applied Solar Energy

OUTDOOR EXHIBITS at the First World Solar Symposium held in Phoenix, Arizona, November 1955. The theme was "The Sun At Work," and solar devices of all kinds from all over the world were displayed.

fuel are scarce, solar energy may be used for cooking and thus save valuable fertilizers now being burned. UNESCO and other agencies are promoting the use of simple solar stoves in lands where there is an abundance of sunshine.

Along with hunger there is thirst. Underground water, stored over long ages of time, is being depleted rapidly and wells are drying up. Sea water, made drinkable by solar distillation, is a possible solution to this problem.

One solar scientist has made the statement that while man does not live by kilowatts alone there is much hope for the prosperity of the world in the use of solar energy. This is particularly true in depressed areas. Israel is an example, and Jerusalem's recent exposition, "Conquest of the Desert," was a glimpse of how the wasteland may someday blossom. India has produced thousands of solar stoves, and in Japan algae are cultivated with solar heat to be used as a food supplement.

Obviously it is not as simple to use solar energy as it is to burn a log. John Ericsson, famous as the builder of the *Monitor* in the Civil War, spent twenty years experimenting with solar engines. In the end he sadly stated that such power was ten times as costly as other methods.

Ericsson and the others did not have the benefit of recent advances in allied sciences, and today the picture is far brighter. Researchers in more than thirty countries are busy, and they profit from the help of an organization called The Association For Applied Solar Energy. Formed in 1954, this group held the first World Symposium on solar energy, in Arizona. The theme of the affair was "The Sun At Work," and the many displays lived up to that title.

Solar furnaces melted high-temperature materials like

Association For Applied Solar Energy

THE LARGEST SOLAR furnace in the world. This curved mirror, designed by Dr. Felix Trombe, was built in an old fort in the Pyrenees Mountains of southern France. Generating nearly 3,300 degrees C., it is used in research and also to melt tons of metal and other material for industry. The reflector, composed of 3,500 small mirrors, is 35 feet across.

platinum, and solar engines pumped water for irrigation purposes. The sun powered radios, telephones and clocks. It generated steam and fired jewelry kilns, besides cooking meals. The Symposium was the merest hint of what is being done, right now, in the field of solar energy.

In France's Pyrenees mountains, a huge solar furnace, thirty-five feet across, melts hundred-pound chunks of iron to produce the purest alloys known. Russia has a furnace of about the same size, and also a solar machine that produces 600 pounds of ice a day! Algeria has a large furnace which can make fertilizer by fixation of nitrogen in the atmosphere.

Homes have long been heated by the sun, but now scientists are beginning to refrigerate them as well. Solar "batteries" have powered telephone lines and can store energy for nighttime requirements. One scientist foresees solar collectors installed along the right-of-way on railroad lines to provide electricity for driving the train. And promising work has been done in the breakdown of water into hydrogen and oxygen, simply by exposure to sunlight. The gases thus obtained are excellent fuels.

Truly, the sun has been put to work, even though the harness is a light one as yet. Each day Sol rises patiently, to shower us with far more power than we need, if we can only learn how to use it efficiently. The task won't be easy, but it must be done. Our scientists know this, and so do the men of industry. Our schools know it too, and quite probably we will some day have men trained especially as solar scientists—"heliotechnologists."

These things are fact, not science-fiction. With each chapter of this book, we will explore a particular phase of solar energy. We will discuss the possibilities and the difficulties, for there are many of both. All over the world, intelligent men are looking to the sun. The same rays strike American and Japanese, Russian and Frenchman. The same sun that cooks a meal for an Egyptian camel driver will produce electricity in a solar battery in Australia.

At present the atom seems geared to the task of destruction, while the sun offers power suited for peaceful tasks. If there is one international possession that can help tie us together, it is the sun. In the years to come, that will be increasingly important to us all.

CHAPTER
TWO

Myth, Legend, and History

Since the sun is the center of our solar system, and such a spectacular one at that, it is easy to understand why man has attributed supernatural power to the sun. From the time he began to think, he turned to the flaming ball of fire in the heavens to explain the unexplainable.

Ra, the early Egyptian Sun-god, was probably derived from an even earlier West Asian deity. The cult of Ra was all-powerful for a long time, and at Heliopolis, the sun capital, there was a sacred tree of Ra, and even a pool where he bathed each evening! Ra had created himself from Nun, the void, and when he tired of ruling Egypt, he ascended to heaven on the back of a celestial sacred cow. He remains there to this day.

Mythology of every land treated the sun as supreme. Apollo was the Sun-god of both the Greeks and Romans. When his son, Phaethon, tried to drive the fiery chariot across the sky, he came too near the earth and catastrophe resulted.

Medea and Circe were thought to be daughters of the

sun, and the sun was the basis for the Golden Fleece Jason sought. In Norse mythology Balder's death in midwinter represents the annual "death" of the sun. Even in the comparatively recent Old Testament, Joshua was buried in the Sun-Palace. Job was told to look at the Sun "when he shined," and *Beth-shemesh* means House of Sun.

The sun was the deity of the Incas. Huge temples were erected in its name at the capital of Cuzco. The Sun-god was symbolized by great discs of gold, and this design was continued in later years by Mexican sun-worshippers. Natives of Brazil thought of the sun as a great spirit who could assume human shape at will.

Our own North American Indians, including the Natchez, Crow, Pueblo, Apache, Navajo, Papago, Cherokee, Iroquois, Blackfoot and Sioux regarded the sun as a chief deity. A legend of the Yuchi tells this charming story of the spirit, On-coye-to.

In the beginning there was no land, only vast dark regions of water. On-coye-to hovered above the water until exhausted, and then dropped like a white feather to its surface. As he did, a whirlwind spun him about on the waves so rapidly that the foam became land for On-coye-to to rest upon.

On-coye-to, though safe, still had no light, until wistfully he saw a distant star, called "Po-ko-lil-ey." He resolved then to bring light to the earth. Visiting the star he found it inhabited by a race of beautiful people. On-coye-to was welcomed everywhere except into the "sweathouse" which was the secret of the wonderful light.

The sweathouse could be visited only by the sick, since the light had magical curative powers. On the day of a big hunt, On-coye-to pretended to be ill and was permitted to enter the sweathouse. Inside were many baskets, each holding a small, bright sun. On-coye-to stole one of the baskets and fled back to earth, hotly pursued by the star-dwellers.

Reaching his own land safely, he hung the sun in its basket up in the sky to light all the earth. Unfortunately the light was not quite right, and On-coye-to has been trying it in different parts of the sky ever since!

Another age-old theme is the "sun-snarer." A young Indian brave lay down to sleep in the noon sun, and his new jacket was shrunk by its heat. Angry, he fashioned a noose and fastened the sun in the sky. The earth soon grew unbearably hot, and in panic all the animals attempted to gnaw the bonds and free the sun. At last the mouse, then the biggest of the animals, succeeded. The sun moved again, but the mouse was reduced to his present tiny size from the heat he suffered.

Eskimo mythology tells of two young lovers who discovered to their sorrow that they were really brother and sister. The girl fled bitterly into the sky to become the sun, and the boy was changed into the moon. The Ojibwa tribe has a similar legend, and also celebrates the "white dog feast," since it is believed the sun and moon eat this meat in heaven.

A tale handed down in Polynesian folklore describes the sun as having once sped across the sky with terrific speed. A young native, angered because of the shortness of the day, solved the problem by simply breaking the sun's leg! With the sun proceeding across the heavens at a more leisurely rate, men could complete their tasks.

So deep-rooted are these myths and legends that they have persisted throughout long centuries. It is said that when Eskimo children play "cat's-cradle" or cup-and-ball, they speak even now of snaring the sun again. Indian sun-dances are still common, and lest we smile in a superior way, don't we call our religious day *Sun*day?

Along with devout worshippers of the sun there were also those who irreverently dreamed of putting the sun to work. These dreamers in time became inventors, and the burning glass was a result of their imagination. The Athenian, Aristophanes, was born in 448 B.C. and mentions the device in his writing. In his comedy, "The Clouds," a character explains how he gets the better of men to whom he owes money. The trick is simply to aim a burning glass at the wax tablet whereon the debt is recorded; a unique way of wiping out an obligation!

The old story of Daedalus and Icarus tells of one of the first aviators coming to grief because of the sun which melted his wax wings. Archimedes, however, turned the sun to his own advantage and used it to rout the enemy. Way back in 214 B.C. the man so famous for levers and other scientific principles was first to use solar energy as a weapon of war. When a Roman fleet attacked Syracuse, Archimedes used "a burning glass composed of small square mirrors, moving every way on hinges—so as to reduce the fleet to ashes at a distance of a bow shot!"

Apparently the Romans profited by the experience in an unusual way for in 77 A.D. Pliny wrote of the use of the burning glass to light fires and also in the field of medicine to burn away dead flesh. Later, in the siege of Constantinople, another Greek by the name of Proclus duplicated Archimedes' triumph by burning an enemy fleet. As usual, the Greeks had a word for it. Helios, an alternate to

Apollo, was their name for the sun. It is represented in our language by *heliograph*, *helium* and many other words.

After Proclus burned the enemy fleet, there was a long lapse in recorded solar history. Eclipses occasionally made the superstitious tremble in their boots, and it is not until 1615 that we find the first solar engine constructed. A man named de Caux was the inventor of this device that pumped water. The engine worked on the principle of air expanded by the sun's heat, and the amazing thing about it is that it was built a hundred and fifty years before Watt built his steam engine.

In 1695 two Italians, Averoni and Targioni, succeeded in "burning" a diamond, using a large magnifying lens. English and German lensmakers were making large glasses by this time and the French Duc d'Orleans used a German lens 33 inches in diameter in many experiments. Oddly, the solar furnace was used to "prove" the erroneous "phlogiston" theory of combustion and it was years before the mistake was rectified.

It remained for a French naturalist, Georges Buffon, to show once and for all that Archimedes' feat was not mere legend. Using a number of small mirrors, Buffon burned wood at a distance of 200 feet, which is about a "bow-shot." He also melted lead from 130 feet, and silver at a distance of half that much. Buffon, along with Cassini, was the first to curve mirrors into the parabolic shape used in today's solar furnaces.

Louis XIV was called "The Sun-King" and rightly so. During his reign much solar research was carried out. M. Antoine Lavoisier was the foremost solar scientist of

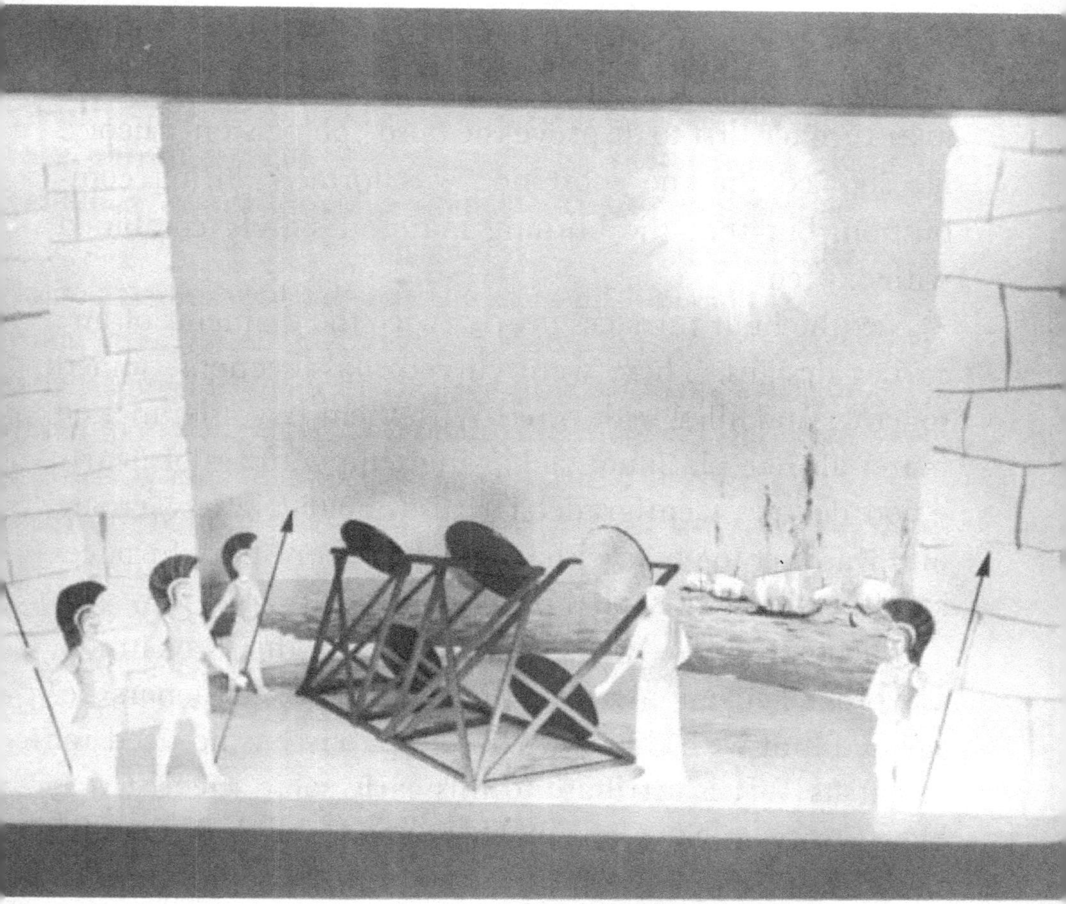

Association For Applied Solar Energy

FIRST MILITARY USE of solar energy! With a mirror arrangement of this type, Archimedes beat off an attacking Roman fleet by setting fire to ships' sails. This took place in the year 214 B.C., and the device ignited the sails "at the distance of a bow shot."

the next generation, and for a time it seemed France would continue to honor researchers. Lavoisier used the solar furnace first to disprove the false "phlogiston" theory. He showed that no substance was formed during combustion, but that the burning material simply combined with oxygen.

Next he built furnaces nearly twice the diameter of any earlier designs. These were curved glass sections, joined together and filled with wine! With them Lavoisier melted materials like platinum, achieving temperatures of nearly 2,000 degrees Centigrade. He also melted in a vacuum, and was first to observe that "the fire of ordinary furnaces seems less pure than that of the sun."

Then, abruptly, with the coming of the Revolution, Lavoisier became a martyr for science. Executioners beheaded him with the cry, "The Republic has no need for scientists!" It is a tribute to this early solar scientist that the factory at St. Gobain which made his lenses, today makes mirrors for the largest solar furnace in the world—in France. And a prime advantage cited by experts is that solar heat *is* "pure," as Lavoisier noted more than 150 years ago.

Men continued to dream of fantastic things. Cyrano de Bergerac in his "L'Histoire Comique des Etats et Empires du Soleil" describes a spaceship that journeys to the sun. It is powered by solar energy. Before we laugh too loudly at Cyrano, consider that two serious-minded scientists recently suggested spaceships powered by—solar energy!

We have mentioned the solar engine designed and built by de Caux, commenting that this was the first such device before Watt's steam engine. Another was produced

Association For Applied Solar Energy

HERE, IN MODEL FORM, is M. Lavoisier's lens-type solar furnace with which he melted platinum and other high-temperature materials and also disproved the erroneous theory that a substance called "phlogiston" was formed by the process of burning. Lavoisier was beheaded during the French Revolution because "the Republic has no need of scientists!"

in 1827, and by the latter half of that century solar engines began to appear steadily.

Mouchot built a solar pump and a water-distiller. John Ericsson, an American by adoption, designed a solar steam engine and several hot-air engines. Built in New York, the largest of these delivered one horsepower for each 100 feet of solar collector area.

Bombay, India, got its first solar steam engine in 1876, a power plant developing 2½ horsepower. In 1880 for the first time solar energy literally made the news when a Frenchman operated his printing press with a solar engine!

At the turn of the century there was a flurry of solar activity in the western United States. A pioneer researcher built huge, cone-shaped reflectors in Pasadena, California, and in Tempe and Willcox, Arizona. These had a collecting area of 700 square feet and were used to pump water. A search is being made in Arizona to locate one built there, for it is believed stored away somewhere.

In 1901 the *Scientific American* described a "band of Boston capitalists" who built a solar engine on the Pasadena Ostrich Farm. Far from being men to hide their heads in the sand, this group produced a device developing 11 horsepower and capable of pumping 1400 gallons of water a minute. It was their hope "that such an engine would result in arid lands blooming as the rose." The one flaw in their plan was that they were years ahead of their time.

California, the land of sunshine always, saw another successful solar engine take shape at Needles in 1905. This developed 20 horsepower, but the builders admitted the cost of construction was about four times that of a conventional plant.

Association For Applied Solar Energy

JOHN ERICSSON'S "Caloric Engine," evidence of the genius of the man who built the famous *Monitor* during the Civil War. Constructed in 1883, this device concentrated the sun's rays with the curved reflector and used the hot air produced to operate an engine.

This same economic truth was learned by The Eastern Sun Power Company, Limited. This organization built the largest solar power plant ever; one with 100 horse-power capacity near Cairo, Egypt. While it operated successfully, it was abandoned during World War I because it could not compete with fuel-fired systems.

Early in this century Dr. C. G. Abbot produced solar power to broadcast a nationwide radio program. In New Mexico a solar pump lifted water into a large tank, which in turn released it to generate electricity which lighted a mine for many years. In Europe and Asia and elsewhere in the world solar heat has long evaporated the water in shallow pools to leave salt. This idea has been taken a step farther and used to purify salt or brackish water.

Eighty-five years ago a huge solar still, covering 51,000 square feet, was built in Las Salinas, Chile. With the sun as the only fuel, this device provided 5,000 gallons of pure water a day, distilled from the brackish supply on hand. Man had come a long way from the ignorant, quaking savage who sacrificed humans on the altar to the sun.

For centuries the sun has parched corn and dried meat for man to eat. In 1837, an inventor named Herschel found a way to step up the heat from the sun and was able to cook vegetables and stew meat in a wooden solar stove. Samuel Pierpont Langley, better known as the builder of what was nearly the first airplane to fly, constructed a solar oven too. In 1884 he lugged it to the top of Mt. Whitney and cooked a meal there.

Since the solar stove is perhaps the simplest application of sun heat, we shall begin our discussion of modern solar devices with it. In the next chapter we learn about different types of solar stoves.

Association For Applied Solar Energy

HERE, AT THE Solar Energy Symposium in 1955, a modern-day scientist duplicates Archimedes feat of using mirrors to start fire at a distance. Charred wood in the foreground is proof of the concentrating effect of mirrors focused on their target.

CHAPTER
THREE

Solar Cookers

Nearly everyone has used a magnifying lens for a burning glass, perhaps even to start a campfire for cooking. Some people have dried food in the sun, and this is a slow form of solar cookery. Even publicity-seekers who fry eggs on the scorching pavement in July are using a crude sun stove! But few of us have ever used an efficient, fast-acting reflector stove or oven, simply because we have so many other fuels available to us.

Fortunately wood is still plentiful in our country, and we can also use charcoal or liquid fuel in camp stoves. It is interesting, however, to consider the advantages of solar cooking. No matches are needed to light the sun stove and there is no flame to create a fire hazard. Carrying along, or gathering fuel on the spot, is not necessary. Solar heat is free, and without smoke, soot or ashes. Add to all this the thrill of cooking a meal with energy literally picked from the air, and you have the essence of solar cookery.

Recently a group of skiers climbed the mountains near Denver, Colorado, in mid-winter. Knee-deep in snow, they

Black Star

SKIERS COOKING lunch high in snowy mountains of Colorado. The solar stove called an "Umbroiler," was invented by Dr. George Löf. It is collapsible for easy carrying, and the silvered inner surface concentrates the sun's rays to produce heat for cooking.

set up a folding stove that looked suspiciously like an umbrella with its inner surface silvered and a small grill attached to the handle. Then, without matches, fuel, or flames, they cooked a meal for the party.

This ultra-modern solar stove is far removed from the earliest of sun cookers but based on the same principle used by pioneers more than a hundred years ago.

In the last chapter, the Herschel solar cooker was mentioned. Built in 1837, this mahogany box, fitted with a glass panel and buried in the sand, cooked vegetables and stewed meat for its inventor. Simple in principle and construction, it generated a temperature of 240 degrees and was called a "hotbox" type of cooker. We are all familiar with the greenhouse, the glass-surfaced structure which lets in the rays of the sun and traps them inside. Herschel's cooker functioned in this manner, heat coming through the glass but unable to escape through the one-way "valve" of the glass partition.

The main disadvantage to this first-recorded cooker was the fact that it could not be moved easily, and while it could have been easily buried at the beach, a picnic on hard ground would have posed a problem! Samuel Langley overcame this difficulty with his better insulated hotbox cooker in 1882. To prove how handy it was to take about, Langley carried it with him on a hiking trip to the top of Mt. Whitney in 1884 and cooked his meals.

Both these solar cookers used only the direct rays of the sun. The first model to use mirrors to concentrate solar heat was a more efficient design, an eight-sided, tapered box a little over two feet across at the large end. This cooked vegetables and meat for its inventor in Bombay, India, in 1876.

Association For Applied Solar Energy

SOLAR STOVES in action. These lightweight aluminum cookers were invented by Dr. Adnan Tarcici, delegate to the U. N. from Lebanon. Made in sections, the reflector can be folded and packed in a small case. To be operated, the stove is faced so that the sun's rays are reflected onto the bottom of the cooking pan. Temperature at the focal point is greater than 1,000 degrees!

Dr. Abbot, whom we have mentioned before, also built a solar stove using mirrors to concentrate heat. Thinking that cooking out-of-doors might not always be practical, he designed an oven with a remote heating unit. Besides being more convenient, this type was not subject to cooling by winds as is the case where the oven itself is placed in the sun.

Obviously, the one big difficulty with a solar stove is that the sun does not shine all day. Abbot realized this, and improved his oven to the point where it would store heat and cook twenty-four hours a day. To prove it could be done, he baked biscuits in a small oven at any hour.

But why should people consider solar stoves when other types were more convenient, and fuels readily available? Even in lands where fuel was becoming scarce, a phenomenon called *inertia* tended to keep natives doing as their fathers and grandfathers had done before them.

Today, however, the solar stove is coming to be thought of not merely as a clever gadget but a useful piece of equipment in places like India, Egypt, and Japan. Besides being short of fuel, India is lacking in food as well. Unfortunately, manure that might better fertilize the soil is used as fuel. UNESCO is much concerned with this, and has done work in the field of solar cooking.

One researcher has designed and built ovens which might be practical for depressed areas. Of metal, with hinged reflectors to focus the sun's rays inside, this oven also stores heat in cans of Glauber's salt, a material which can hold great quantities of heat. After sunset the salt releases this heat and continues to cook. Thus the evening meal may be prepared after dark.

This oven has been demonstrated countless times and

ALUMINUM SOLAR COOKER designed by M. L. Ghai for the National Physical Laboratory of India. This type stove is mass produced in factories in India, and sold for about fourteen dollars. The side of this cooker is cut on a straight line so that it may be used even in the early morning and late at night when the sun is very low in the sky.

has proved itself in the field. Compact, and generating up to 400 degrees of heat, it can be mass-produced for as little as five dollars. This should make it a boon to natives in India, for though fuel is dangerously scarce, sunshine is not. Surveys show that the sun shines for an average of 310 days a year. This means hot meals about six days each week.

We need not go as far as India, either, to find a need for such a device as the solar stove. A speaker at the first World Symposium on Solar Energy described a village in Mexico where the women walk six miles into the hills for wood to cook with. Paradoxically, there is more than an abundance of sunshine the year round in that location to cook with, heat, and even cool homes when needed!

So far we have talked mostly of "hotbox" or oven-type solar stoves. The skier's stove mentioned earlier is another kind of cooker—a reflector. Instead of trapping warm air inside a closed box, the reflector simply focuses the rays from a large area onto a small spot to heat a cooking vessel of some sort.

In Japan where rice is the staple diet, it is not surprising to find solar rice-cookers. One popular type is composed of numerous small, flat mirrors, arranged to form a cone. Reflected sunlight strikes the blackened surface of the rice pot and cooks the cereal inside.

Just as a properly shaped lens makes all rays passing through it converge at a focal point, a reflector concentrates a large quantity of *warmth* to make a small amount of *heat*. The surface of the reflector is curved, so that the sun's rays strike it at an angle and rebound toward a common center point.

Association For Applied Solar Energy

RICE COOKER manufactured by the Goto Company of Japan. Flat mirrors are mounted inside a wooden framework and adjusted so that they reflect the sun's rays onto the blackened cooking pot. This cooker will boil a quart of water in about 30 minutes. Solar stoves are especially useful in countries like Japan where fuel is scarce.

To make these rays meet in the tiniest point possible requires a parabolic shape, which is a specially computed mathematical curve. Searchlight reflectors and high-temperature solar furnaces must be made in this manner and are very expensive. However, fortunately for those who would cook with sun heat, a spherical shape is good enough for all practical purposes. In fact it is even better than the parabola, since such a perfect reflector would develop enough heat to burn a hole in the bottom of the cooking vessel!

Using the spherical, or perfectly round reflector, gives a larger spot of lower temperature. The simplest solar stove of this type is one made at the University of Wisconsin. It is literally scooped from the sand, a sort of modern echo of Herschel's earlier hotbox cooker!

A tripod is set up over the spot where a cooker is desired, and a calculated distance from the ground. A string is hung from the tripod with a curved scraper attached to its end. With this scraper a depression is scooped from the sand or soil, the result being a spherical hollow. A thin coating of cement is laid over this and then aluminum foil applied to form the reflecting surface.

Ideally the hole is scooped in a south-facing hillside with a slope of about 15 or 20 degrees so that the reflector will be approximately at right angles to the sun's rays. A post is sunk in the center of the depression and a grill attached to hold the cooking vessel. The grill, of course, is placed at the focal point. This is easily calculated since it is half the radius of the sphere.

Crude as this reflector seems, it functions and will cook a meal. Such a solar stove costs next to nothing and has been suggested as a possibility for inhabitants of depressed

University of Wisconsin

A SIMPLE REFLECTOR stove scooped from earth and lined with foil. This is one of many low-priced cookers developed by Dr. John Duffie and other researchers at University of Wisconsin. The depression was formed by suspending a wooden paddle above the soil, using a string of the correct length. Such a stove is not very portable and is best suited to cooking the noonday meal.

areas. There is one limitation; it will operate only at one time of day, say for cooking the noon meal.

The University of Wisconsin has experimented with a number of other types of solar reflector cookers which cost from five to nine dollars for material. One design uses fibreglas, molded over a form in much the manner that boat bodies are. The resulting shell is lightweight and durable. Its inner surface is lined with reflecting material and fitted to a suitable base. Such a cooker would be a good project for scouts or home craftsmen.

One unusual cooker is made of inflated plastic material. It is believed that this design has possibilities for mass production, and besides being cheap it would be easily portable. For all its flimsy appearance, such a cooker will boil a quart of water in twenty minutes or so and cook food in proportionate times.

The average reflector stove is about forty inches across, and not too difficult to carry around. These have cooked everything from bacon and eggs to chicken. All that is necessary is to adjust the reflector so that the hot spot falls on the bottom of the cooking vessel. Every 15 minutes or so it will be necessary to move the reflector slightly to follow the sun. And the cooker will function from sunup to sunset.

As mentioned before, a blackened cooking pot will be more efficient, since black absorbs instead of reflecting the heat from the cooker. While it is possible to fry meat handily, a pressure cooker makes a better companion piece for the solar reflector stove because it does not let the heat escape into the open air. One word of caution in case you come into possession of a solar cooker: The

Association For Applied Solar Energy

DR. MARIA TELKES demonstrates her solar oven by preparing hamburgers. The extended reflectors send sunlight through the glass top of the oven. Temperature inside is 450 degrees, and this type of solar stove can store heat for cooking after dark.

focal point may be in the neighborhood of 1,000 degrees so be careful!

We have talked of the folding, umbrella type cooker and it is obvious that such a device is easier to pack than a rigid aluminum or fibre shell. A Lebanese delegate to the U.N. has designed a folding aluminum cooker that goes into a small container. This feature does not affect the performance of the cooker, and it has fried eggs in Central Park in December. The stove has been produced in Egypt, and its inventor feels that mass production would permit a price of about ten dollars.

A solar researcher in India designed a compact one-piece stove in 1952 and had it produced by a factory in quantity. It cost fourteen dollars, which unfortunately is a large sum of money in India. Besides this disadvantage, there are two others. One is the cooling effect of the wind in cooking outdoors, and the other is the task of educating the Indian people in the use of such a modern device.

This is the solar stove of today, novelty to some and possible blessing to many others. Researchers at the University of Wisconsin have set themselves a goal of a cooker developing 500 watts at a price of not more than five dollars. These devices can make it possible for nomadic peoples in the world's wastelands to prepare hot meals and beverages where it would be otherwise impossible.

There are men who foresee the day when solar energy collectors on the roofs of our homes will cook our meals in remotely heated, built-in ovens. When that day comes we may smile to recall the "old days" when cooking with gas was really something!

CHAPTER
FOUR

Solar Furnaces

In the preceding chapter we talked about sun-powered cookers. Now we come to the solar stove's big brother—the solar furnace. This device, reminiscent of the fiery furnace of Biblical times, is the most spectacular use of the sun's heat.

Magicians use mirrors to create illusions, but solar scientists produce an even greater magic. Today, shiny curved mirrors are doing research and even industrial chores like melting metals previously unmeltable, and creating fertilizer from nothing more than pure air and sunlight!

Remembering that the surface of the sun is a terrific 6,000 degrees C., it is indeed fortunate that such heat does not reach the earth. The record temperature for the most sweltering day is a mere 125 degrees or so. However, by concentrating the sun's rays on a smaller area, we can increase the temperature. With a magnifying glass enough heat can be generated to start a camper's fire or carry out a prankster's joke.

The solar furnace is a refinement of previously mentioned devices. If lenses are used, these are of highest optical quality. Curved mirrors, instead of being spherical, are in the shape of a paraboloid. This "perfect" shape focuses the gathered heat into the tiniest focal point with the result that very high temperatures are available.

Since no instrument is truly perfect, the sun furnace does not quite match the heat of the sun's surface which is the absolute limit in theory. Temperatures reached are amazing, nevertheless. Already scientists have recorded higher readings than the oxy-acetylene torch, and they believe that with better furnaces temperatures perhaps as high as 4,000 C. can be reached!

Reviewing feats of earlier solar researchers, we discover that Cassini melted iron and silver more than two hundred years ago. Diamonds had been melted, and Lavoisier achieved more than 1500 degrees C. These men did wonderful things with the tools at hand, but it was not until this century that a really high temperature solar furnace was built.

In 1921, a modern furnace was made in Germany using silvered-glass mirrors. The precisely ground reflector produced a temperature of more than 3,000 degrees C. This introduced something new by using an auxiliary mirror to reflect the sun onto the parabolic mirror which was fixed for easier handling of experiments. The auxiliary mirror, called a heliostat, tracked the sun as it moved across the sky.

California Institute of Technology built a fine solar furnace in 1932. Instead of using a single mirror the California furnace employed nineteen lenses, each two feet in diameter. This array of lenses gives a total collecting surface of nearly sixty feet, and temperatures higher than 3100 degrees C. have been reached.

Association For Applied Solar Energy

SOLAR FURNACE operated by California Institute of Technology. One of the first furnaces built in this country, it is used in the high-temperature research program of California Institute of Technology.

The first industrial use of the solar furnace came in 1940 with a ten foot mirror built for the AC Spark Plug Division of General Motors. Since optical glass was critically short at that time due to World War II, the mirror was spun from ¼ inch thick aluminum and then polished. Also developed was a sliding tube which controlled the temperature at the focal point of the furnace. At present this furnace, the largest in the country, is being used for research in high-temperature aircraft materials.

In our modern world there is a growing need for higher temperatures. The fields of metallurgy and chemical synthetics are good examples. It is therefore understandable that industry is much interested in the possibilities of the solar furnace.

There are ways of reaching very high temperatures with arc furnaces, electrical and electronic methods, and the experimental "shock wave" technique. However some of these yield their peak temperature for only a split second. One arc device which has been suggested for industrial use has an efficiency of only 1%, besides the fact that the electrodes create magnetic and electric fields. They also get in the way, and contaminate the metal being melted.

The solar furnace, on the other hand, can be sustained all day long or at least while the sun shines! There is no soot, no ash, no harmful electric, magnetic, or chemical action. A solar furnace is extremely safe to operate, and for a final reason the power supply is free for the taking. Almost two hundred years ago Lavoisier noted prophetically these advantages which have stimulated current interest in the furnace. Fortunately, today republics recognize the need for scientists.

Builders of high-speed aircraft and missiles are vitally

United States Air Force

SOLAR FURNACE MODEL built by the United States Air Force. At right is a movable mirror which follows the sun as it crosses the sky. On the left is the parabolic mirror which focuses the sun's heat into the "furnace room" hidden back of the mirror. The venetian blind arrangement in the middle is called an "attenuator," and serves to regulate the light admitted to the curved mirror. This controls the heat at the focal point.

concerned with high-temperature research. With airplanes approaching the 2000-mile-per-hour mark and huge missiles traveling at speeds more than double that, the so-called "thermal barrier" poses a problem engineers must overcome.

Many materials have been tried in the solar furnace and found wanting. At temperatures achieved only in the furnace—or in actual flight conditions!—wing leading edges and other parts explode, melt, and even simply evaporate. Then engineers must come up with a substitute that will stand the strain of the heat barrier. This same sort of work is being done in the field of "cer-mets," or materials composed of metal and ceramic bonded together.

Besides the aircraft corporations, many other organizations are already using solar furnaces. Kennecott Copper Corporation has installed a small one at its works in Salt Lake City. General Electric has used a solar furnace to purify phosphor for communications equipment.

The Government's Bureau of Mines maintains a furnace in Morgantown, W. Va., and the Sandia Corporation uses one in high-temperature research work. The Army's Quartermaster Corps is building a large solar furnace, and the Association For Applied Solar Energy intends to erect one in Arizona.

For some time schools have been interested in the furnace as a research tool. California Institute of Technology has been mentioned; there are many others. Fordham University operates a furnace, and MIT uses one to study the effects of radiation on materials. At Arizona State College at Tempe, the first heliostat-equipped furnace in this country has been installed. It is doing research under government contract.

Association For Applied Solar Energy

THIS FIVE-FOOT Convair Solar Furnace, being demonstrated by B. I. Davis, a thermodynamics engineer with Convair, San Diego, California, is constructed of a surplus searchlight mirror, and is capable of attaining very high temperatures in the neighborhood of 3,000 degrees C.

All around the world, everywhere the sun shines, there is interest in solar furnaces. Australia, India, Russia, Algeria and France are among the countries doing solar furnace work. Symposiums have been held, and one university has offered a course in utilization of solar energy. All this points to the growing importance of the solar researcher, the "heliotechnologist," as he is coming to be called.

The ten-foot solar furnace, though the largest in the United States, is small when compared with some in other countries. Algeria, for example, has a furnace approximately 26 feet in diameter. Besides doing research in the high-temperature field, this single-mirror furnace makes fertilizer by the process of fixation of nitrogen from the atmosphere! Thus the sun not only makes crops grow in Algeria, it fertilizes them as well.

Russia has a large furnace, perhaps over 30 feet in diameter, constructed of reinforced concrete and powering a steam plant. But to France goes the honor of operating the world's largest and most advanced solar furnace. This seems no more than fair, with the long history in sun research France has as a heritage.

High in the Pyrenees mountains, in old fort Mont Louis, Dr. Felix Trombe headed the building of a huge, 35 foot diameter reflector which generates 75 kilowatts of power and melts more than 200 pounds of metal at a time. This is truly an industrial furnace, since for 30 days each year it produces metals and refractory materials for French firms.

Experimenting at first with searchlight mirrors, scientists became convinced that a large furnace would be an

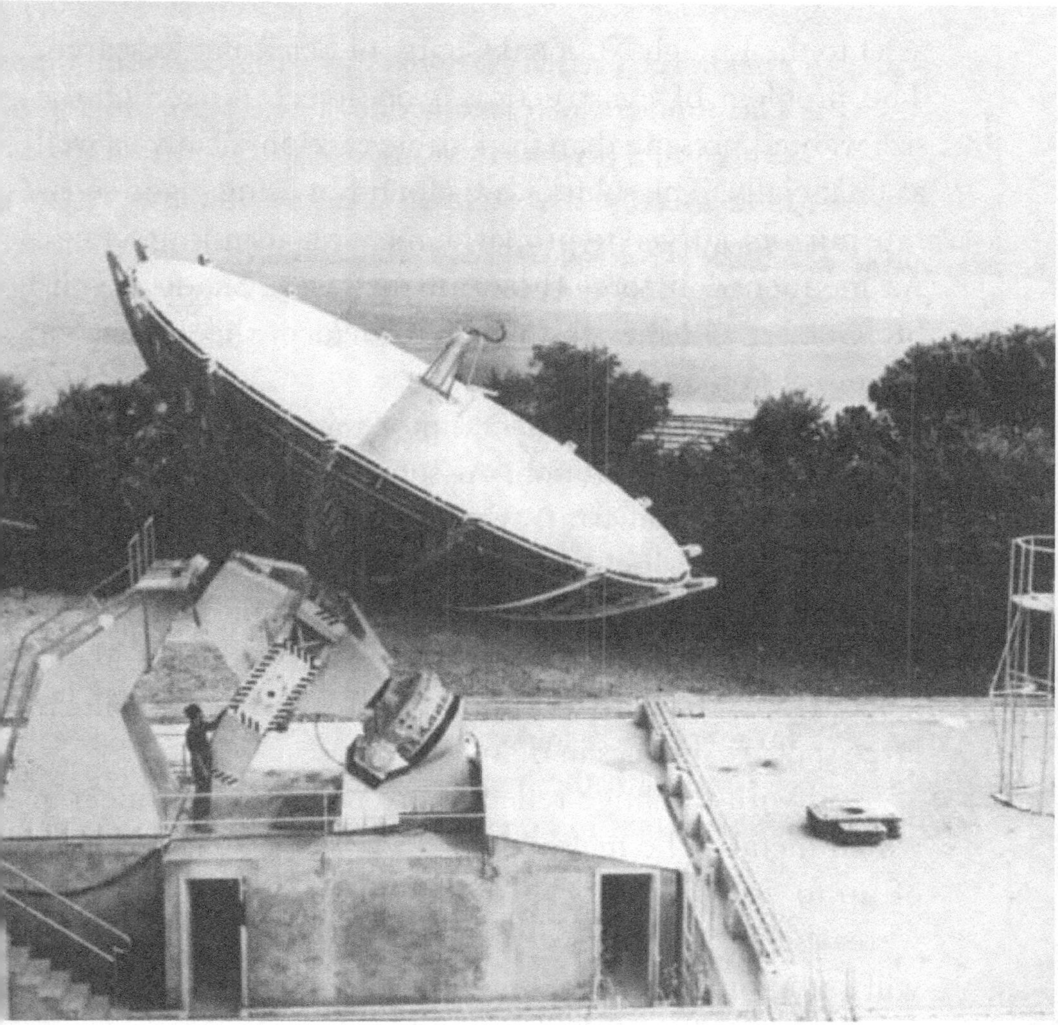

Association For Applied Solar Energy

THIS SOLAR FURNACE, one of the largest in the world today, is operated in Algeria. One novel use for it is the "fixation" of nitrogen from the atmosphere. The nitrogen thus produced is used for fertilizer. Notice the mounting that permits the reflector to turn, instead of being fixed as is the case with France's 35-foot furnace. The furnace shown here needs no auxiliary "heliostat" mirror.

asset to the French National Center of Scientific Research. The problem of constructing a one-piece mirror of the size wanted presented an impossible task physically as well as financially. The solution was found in using 3,500 separate mirrors, curved to proper shape with adjusting screws. As mentioned before, these mirrors were made by the St. Gobain glass works which provided the lenses for Lavoisier so many years ago.

The curved reflector made up of these small mirrors is fixed in a vertical position. At some distance from it is a flat mirror, the heliostat, which moves slowly during the day and thus positions the sun's rays on the curved reflector. Although the small mirrors used are not of the finest optical quality, the furnace achieves temperatures above 3,000 degrees C. The focal point of this furnace is a fairly large spot, 4½ inches in diameter.

Used with the huge reflector are rotating containers in which metal to be melted is placed. These spin at speeds of up to 750 revolutions per minute and are cooled by a water jacket. All this is necessary so that the molten metal will not in turn melt the sides of the container!

Alloys or other materials produced in the Mont Louis furnace are extremely pure, since there is nothing to contaminate them. A technique used successfully is that of placing material to be melted inside a pyrex or quartz flask. It is then melted in a vacuum or gaseous envelope, since the sun's rays pass right through the glass.

Supplementing the large furnace are several smaller models. Some of these are mounted horizontally, which makes it easier to melt powders. A carriage mounted on rails is used for this purpose, slowly advancing the material

CLOSE-UP VIEW of the 35-foot-diameter French solar furnace at Fort Mont Louis. This large reflector is made up of 3,500 small flat mirrors, each bent to a slight curve. Heat gathered by this furnace is equal to 75 kilowatts of power, and melts loads of a hundred pounds of metal or more.

into the focal point. With the curved mirror lying flat there is plenty of "elbow room" for such procedure.

An unusual sidelight at Mont Louis is a freezer operated by the sun. It seems strange that the same hot rays that melt materials such as zirconium also can operate a refrigeration unit, but there is much promise indicated in this direction of solar research.

The solar furnace can melt at three fourths the cost of an equivalent arc furnace, and if used solely for industrial purposes the 35-foot mirror would have paid for itself in five years. Although the furnace obviously functions only during sunlight, it is operated more than 2,750 hours per year. Even when there is snow on the Fort, the furnace draws sufficient heat from the sun.

So successful is this giant that France is building one that will make the Mont Louis furnace seem tiny—a mirror 120 feet across! This great installation will have a power of 1,000 kilowatts, or the equivalent of a 1300 horsepower generator, and Dr. Trombe feels that it will melt metal industrially at half the cost of electric arc methods.

Scientists in our country are so interested in the big French furnace that a team has studied the plans and the work as it progresses. The United States Air Force has announced plans to erect a furnace of the same size—120 feet—at Cloudcroft, high in the New Mexico mountains. Models of this proposed furnace include a shutter interposed between heliostat and curved mirror to regulate the temperature at the focal point.

A research organization has suggested furnaces of 50 to 100-foot size for the iron and steel industry, and perhaps someday the solar foundry will be a reality. The solar fur-

Earl S. Ream Photo

THIS LARGE SOLAR POWER PLANT was built in 1903 in Arizona by A. G. Eneas. Its cone-shaped collector, about 30 feet in diameter and composed of a large number of small mirrors, focused heat on a boiler. The steam generated drove an irrigation pump. Notice the glowing boiler tube made of glass, and also the water pouring from the pipe at lower left.

Association For Applied Solar Energy

THIS LARGE FURNACE, possibly 35 feet across, was built by the Russian Heliotechnical Laboratory in Tashkent. A square tank at the top (not shown here) suggests that this installation is used to generate steam power. Another type of reflector is seen in the lower right background.

nace has already proved it can deliver higher temperatures than any other practical method; it may well invade the general melting field. Though it has the drawbacks of unwieldy size and large initial cost, it has advantages to offset these:

There are no parts to wear out on a solar furnace; the Mont Louis mirrors have withstood six years of weathering and suffered no ill effects even in gales that damaged slate roofs in the nearby town. Even more important, the power is free. As someone aptly quipped, the solar furnace is the hottest thing in the industry!

CHAPTER
FIVE

Solar Air Conditioning

It has been accurately stated that civilized life, as we in this country think of it, would be impossible without house-heating. Anyone who has run out of fuel in midwinter in the northern United States can appreciate this fact. Keeping ourselves warm is so important to us that we use somewhere between one fifth and one third of all our fuel for heating. Each year we dig into our reserve supply the equivalent of nearly one half *billion* tons of coal for this purpose!

Power engineers point out that besides diminishing our stockpile, such use of fuels is wasteful and inefficient. To heat our houses requires only relatively low temperatures; conventional fuels would be much better used in producing high-temperature energy. The sun, however, is the ideal solution to the problem of space-heating, and such an application of solar energy has already proved itself today.

So optimistic are the experts that the President's Materials Policy Commission has predicted the solar-heating of as many as thirteen million dwellings in the United

States by 1975. Even assuming that solar energy could save only half the fuel now spent on heating, this economy would add twenty-five or thirty years to our fuel reserves.

Even in winter, the sun's rays will warm us. Houses with windows facing south gain considerable heat in this way. By very simple means, solar scientists can step up the degree of heat so that the sun can take over the job ordinarily done by gas, oil, or electricity. Since each home can collect its own solar heat, there is no dependence on a central power station or transportation system that can break down just when needed most. Neither is there any loss of energy in transmission, as is the case with power lines.

Solar water-heating is no novelty. For a long time man has recognized this phenomenon and applied it. Rice paddies in the Orient are often irrigated with water allowed to flow in a shallow trench before contacting the plantings. Such warmer water results in better yield. Another example of solar water-heating is the evaporation of shallow ponds to produce salt.

From these simple beginnings it is not surprising that man progressed to heating water for domestic use. The solar water heater was a common sight in California and Florida the early part of this century. At first the tank itself was placed so the sun would strike it and thus slowly warm the contents. Later the idea of coils of pipe made for a faster-heating system. Finally the pipes were blackened to absorb even more heat, and the whole business was enclosed in glass to take advantage of the glass-house principle we have mentioned before.

While these early water heaters were usually poorly engineered, they functioned fairly well. It was only the

Hoffman Electronic Corp.

HOUSE OF TOMORROW. This solar-shingled house, deriving all of its power needs from solar cells on the roof, was forecast as a future application of solar energy by Hoffman Electronics Corporation, Los Angeles. The 40x20 foot roof section, shingled with silicon solar cells, would enable the roof to supply 8.8 kilowatts per hour in sunlight. In this manner, only five days of sunlight per month would be needed to develop the proper energy to operate the home.

inconvenience of having bad weather interrupt the hot water supply, and the cheapness of fuel that caused the wane in popularity of the solar heater.

With the introduction of automatic washers, dishwashing equipment, and the need for more hot water, power consumed in water-heating may now be as much as one fourth of the space-heating requirement. Interest in the solar water heater has revived. In 1951 it was estimated that Miami, Florida alone had thousands of such installations. These ranged in size from fifty gallon capacity in private homes to thousand gallon units in apartment buildings. Even though the engineering was not always the best, these heaters generally paid for themselves in from four to seven years. Besides this is the even more important fact that they saved large amounts of conventional fuel.

Today researchers have devised more efficient collectors and worked out proper designs. A collector about seven feet square will provide sufficient heat for a fifty gallon storage tank, and water passing through seventy feet of coils is raised 30 degrees F. Overnight storage is possible, but in areas where there are days of bad weather, supplementary heating methods are of course necessary.

The town of San Remo, Italy, boasts of a brand-new solar water-heating installation. Work is being done in Australia, Russia and many other countries towards more efficient equipment. Some lands which do not have the technological facilities we enjoy have worked out simpler methods. In India, researchers imbed water pipes in concrete for a much cheaper installation.

Besides being economically competitive with conventional types of water heating, the solar heater offers hot water in places where it would otherwise be unavailable—

remote communities, mountain cabins and the like. Solar research has also devised a better way of warming water in swimming pools! This is done by covering the pool with a transparent plastic which admits light but prevents evaporation losses.

Having the sun heat our water is well and good, but stepping from a hot bath into a frigid house would be little comfort. Let's consider how the sun may also heat the house as well. In the first chapter it was mentioned that the roof receives, and wastes, far more energy than all our appliances use in the form of electricity. Solar space-heating offers an excellent way to put the roof to work as a collector for the heating plant.

Normally, most of the heat energy bombarding the roof is reflected back into the atmosphere. In fact reflective shingles are designed to do just that, to waste heat energy so that our homes will be cooler in summer. Later on we will talk of how this heat can be used to refrigerate the home in summer; right now let's stick to the heating problem.

By replacing part of the usual shingles or other covering with glass-plate collector, the solar engineer "soaks up" sun heat. Air is circulated through the collector and then through a normal duct system to warm the rooms. The temperature of this air varies from about ninety degrees in the morning to as high as two hundred degrees at mid-day. Obviously this latter is far more than is needed, and thermostatic controls keep the house from over-heating. Later we will see what becomes of this surplus heat.

In actual tests it has been found that one hundred per-cent solar heating in a cold climate requires a very large collector, therefore most engineers set themselves a goal

Black Star

LARGE SOLAR water heating installation being completed atop apartment building in San Remo, Italy. Water is piped through coils underneath the special glass plates shown and its temperature is raised as much as 70 degress. A large storage tank will take care of night requirements.

of about three-fourths solar heat, with the balance made up by supplementary heat in the form of gas, oil, or electricity. For this purpose a collector of about six hundred square feet has been found ample for a house with fifteen hundred to two thousand feet of floor space. Either roof or wall can serve as collector, and both types have been used.

A logical question comes to mind when we think of solar heating. What happens when the sun goes down, or when it fails to shine for some time; does the homeowner don his overcoat and mittens? Fortunately this is not necessary, nor must he turn on the gas furnace either. During the day, surplus heat has been stored up. At night, or during cloudy weather, this stored heat is drawn upon to warm the house.

There are a number of ways to store heat. One method is to fill a bin with gravel and pass the surplus warm air over it. After dark, the direction of the airflow is reversed, and the gravel gives back its heat for warming the house! Some designers feel that water is the ideal storage medium and provide a large storage tank which connects with the system. Yet another idea is to use a material which melts when heated to a certain point and stores a large quantity of heat in this manner.

One early solar heated house was constructed by researchers at MIT in 1939. The MIT designers used the water tank storage system. This house, and a second, were purely for testing purposes. However, a third house, built in 1948, has been lived in continuously by student families. Solar heating takes care of three fourths of the needs of the house.

In 1948 another solar-heated house was built at Dover,

Association For Applied Solar Energy

THE DOVER SOLAR heated house. The large second floor "windows" are really heat collectors. During the day surplus heat is stored in containers of Glauber's salt which hold five times more heat than an equivalent volume of water. After dark this stored heat is called on to warm the house. The Dover house gets 95% of its heating from solar energy, despite cold Massachusetts winters.

Massachusetts. One innovation was the use of containers of Glauber's salt for heat storage. This material holds five times as much heat as an equivalent volume of water, and proved the feasibility of solar heating over periods of several cloudy days. The Dover house has furnished up to ninety-five percent of the required heat from solar energy.

Another solar heated house was tested in Boulder, Colorado. A small bungalow, equipped with a roof-top heat collector and a storage bin of gravel, effected a fifty percent saving in fuel. The designer plans a larger home in Denver which employs a collector of six hundred square feet to provide three fourths of the heating.

In Arizona, an architectual firm has constructed a home entirely heated by solar energy. A further development uses part of the same duct and storage system to partially cool the house in summer. The idea of using the sun's heat for cooling in mid-summer is an intriguing one and much work is being done in this direction.

An example of solar energy being used for heating on a large scale is a beautifully modern office building in Albuquerque, New Mexico. This structure has 4,300 square feet of floor space, and one whole wall is formed by glass collector panels.

Heat storage is provided by a 6,000 gallon underground tank in winter. The sun is expected to provide ninety percent of the heat for the building. In the summer time "heat pumps" help to cool the building. An interesting feature is the ability of the system to remove heat from the air during the daytime in spring and fall, store this heat and use it at night to warm the building!

In the scorching heat of summer, solar energy sounds like the last thing we would want. However clever engi-

Rose Photo

FIRST COMPLETELY solar-heated office building in the world. This modern structure was designed and built by the firm of Bridgers & Paxton in Albuquerque, New Mexico. As shown, one entire wall is of glass, backed by metal coils. Water circulates through these coils and is warmed by the sun. Part of it is used to heat the building, and any surplus is stored in a 6,000 gallon water tank buried underground. At night and on cloudy days this surplus heat is used.

neers are working on methods of putting the sun to work operating refrigeration plants! Although this sounds fantastic there is actually a greater potential here than in winter heating.

As a heating source, the sun shines its hottest when it is least needed but the reverse is true in the case of refrigeration. At mid-day, when cooling is most necessary, Sol delivers the greatest power. Efficiently used, this power can keep us comfortably cool in the hottest weather. Thus a six hundred square foot collector atop a home would provide enough refrigeration to cool it in summer.

Strangely, the idea of cooling with solar energy dates back at least as far as 1889. In that year Tellier proposed using a "hotbox" collector to power an ammonia refrigeration system. More recently, in 1938, a scientist suggested burning hydrogen to power a refrigeration system. He would obtain the hydrogen from water decomposed by solar energy.

Besides the mechanical refrigeration plant there is another type known as "absorption cooling." Although not in general use, this absorption technique is excellently suited to solar energy and would require only half the collector area per ton. Thus the six hundred square foot collector which would comfortably warm us in winter would provide up to five or six tons of refrigeration at mid-day. This would be ample cooling for a house in the hot southwest regions of our country.

Experiments have been made with a third type of cooling, a glycol de-humidification process which would be efficient when coupled with a solar energy collector. This type would be most effective in a humid climate, and has

Association For Applied Solar Energy

THIS HOME, belonging to Japanese scientist M. Yanagimachi of Tokyo, is air-conditioned the year round by solar energy. The large screen is a flat plate heat collector, and provides warmth during winter, and power for cooling in summertime.

Association For Applied Solar Energy

RUSSIA, TOO, is active in solar heating experiments. Shown here is a water-heating collector which supplies a bathhouse. Blackened coils can be seen beneath the glass cover. The collector is placed at a slant so that the sun's rays will strike it nearly perpendicularly during the hottest part of the day.

the added advantage of removing unwanted moisture from the air.

While not many of us will move into a completely solar air-conditioned house for some time, it is logical that there will be rapid development in that direction. Architects are aware of the benefits of designs which keep the sun in mind. Improvements are constantly being made in collectors, like the highly-efficient "selective surface" coating developed by Israeli scientists.

Although a solar-heating plant now costs more to install than a conventional type it will break even in about fifteen years. With volume production, costs may possibly be brought down to equal other types.

CHAPTER
SIX

Something New Under the Sun!

So far we have considered the sun principally as a source of energy for stoves, furnaces, and house-heating devices. For years, however, men have cocked an eye toward Sol and dreamed up literally every idea under the sun. Archimedes' solar weapon has been mentioned, as has De Bergerac's space machine, and these only scratch the surface. A glance at patents granted in this country will give an indication of the range of solar inventions.

In 1852, Ulysses Pratt patented a solar ivory-bleacher. Apparently there were huge quantities of ivory needing bleaching, for by 1883 four more patents had been granted for such contraptions! Of course ivory was not the only thing to be dried by the sun.

One early patent, issued in Civil War days, pertained to the curing and drying of fish. Raisins and other fruit also were treated in this manner. As late as 1920 a patent was granted for a solar fruit drier, and more recently, in 1950 as a matter of fact, two patents were allowed for solar treatment of Irish moss.

Association For Applied Solar Energy

THIS MODERN SOLAR CLOCK is a far cry from the sun dial! The collector seen on top of the clock converts light into electricity, driving the motor, and also storing power for use after the sun has set.

SOLAR-POWERED HIGHWAY FLASHER. Continuous operation at night is assured by means of storage cells which hold power obtained from the solar cells during the daytime.

Hoffman Electronic Corp.

Branching out in another direction, one inventor came up with a solar brick-drier in 1883. Applying this technique to another building material a present-day inventor in India has designed a sun-heated lumber kiln. Perhaps one of the most unusual applications of sun driers was that made specifically to bleach feathers way back in 1878!

The year 1881 saw the patenting of a method of warming and ventilating apartments with the sun's heat and this has been a favorite of inventors ever since. Water heaters are popular too. Some researchers were far more advanced, however. As early as 1897 there was a patent granted for the generation of electricity by sun heat on "thermo" batteries. This device was evidently a crude forerunner of the modern solar battery we are familiar with.

Sun pumps and other types of motors date back to 1880 in patent history, but along with these were more homely applications of the energy from the sun. In 1898 a man named Rieke patented the first and apparently the only combination solar stump-burner and insect exterminator!

In 1917 a patent was allowed for a solar solder-heating tool. This interesting use of sun energy is repeated today in an improved device which a home craftsman can easily build and with which he can obtain temperatures as high as 2,000 degrees. There are other applications of the solar "kiln" which we will discuss later on in this chapter.

Although the solar still was known and used in the latter half of the nineteenth century, it wasn't until 1919 that the first American patent was granted. Since then there have been many improvements, including the inflatable plastic stills included in survival kits of aircraft flying over the ocean. The solar still is a simple project, made from inexpensive materials, and can provide distilled

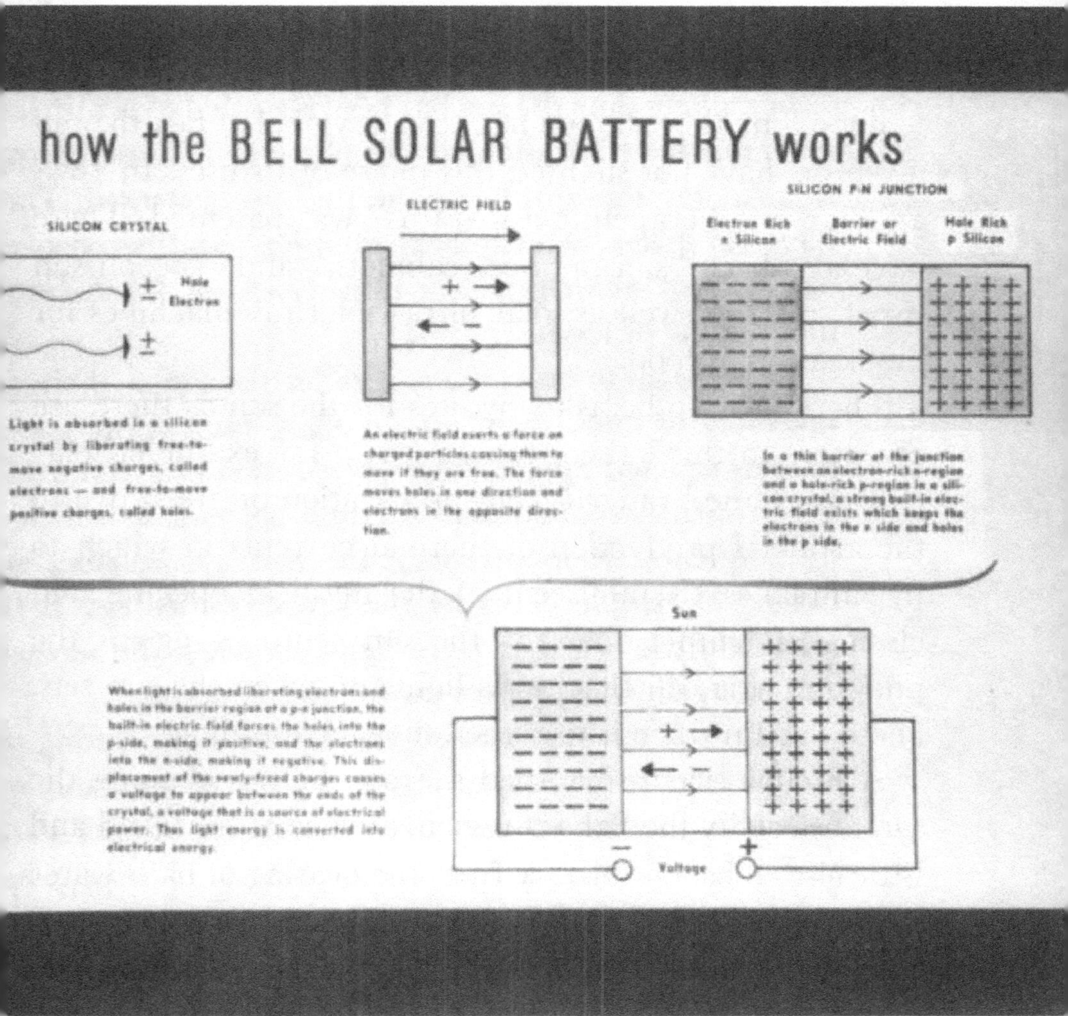

Bell Laboratories

water for a number of uses. Recently a French researcher has suggested a solar *sterilizer* for water.

For a long time the sun has been a symbol of health, and inventors have not slighted this phase of its uses. In 1934 a "solar ray therapeutic apparatus" was patented. There have also been "sun ray heat cabinets," and "cocoon sun sweat suits," as well as solar ultra-violet ray machines for the treating of food.

There seem to be as many uses for the sun as there are inventors to think them up. In 1945, for example, solar patents granted ranged from preparation of green dye to the artificial production of thermal currents in which to fly sailplanes! Reminiscent of the mystical opening and closing of temple doors by the sun centuries ago is the photo-electric cell that turns lights on when the sun sets, and off again when it rises the following morning.

A French scientist studied the possibilities of using the sun to electrify the Sahara region, a place of sun, sand and little else. Jobs for Sol vary from the heating of bath water at Colomb-Bechar, a fact accomplished, to a proposed trans-desert railroad powered by solar-generated electricity. Other Frenchmen believe the only sensible approach to using solar energy is that of somehow tapping the difference in temperature of surface ocean water and water at a depth. Although this is far afield from what we generally consider solar energy, it is the sun of course that heats the sea.

While an Indian factory produces solar cookers, an American manufacturer is seriously studying the possibility of a sun-powered motor for boats! For the outdoor man there is also the solar radio. This is already on the market, equipped with a small solar battery instead of a

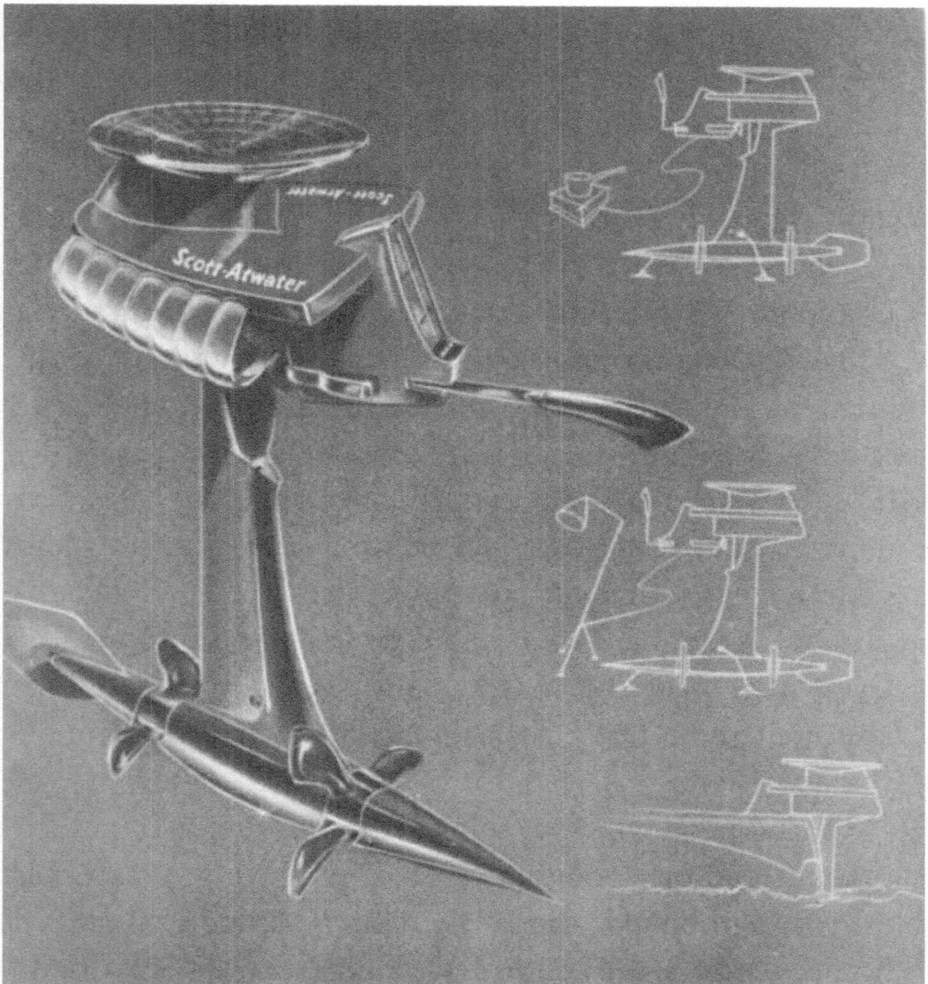

Scott-Atwater Mfg. Co., Inc.

THE SUN WOULD provide the fuel for this solar "dream" outboard motor, conceived by a leading manufacturer. Sun's rays would be trapped by the concave disk and converted to energy that could be stored for long periods. Not only could the solar motor be used for powering a boat, but also for operating camping equipment.

conventional power source, and able to store up power during sunlight so that it can play after dark.

Time and the sun are almost one and the same. In the beginning, man simply divided the day into light and dark. Gradually he learned to tell time more accurately with a sundial. Today there is a solar clock, made in Switzerland, which tells time to the second and needs no winding. Using only sunlight, the solar clock will run continually.

Besides heating homes, the sun has been made a farm-hand by doing the same for barns and chicken houses. Solar pumps can provide water, and solar stills purify it if it is brackish. Fertilizer literally "from the sun" has been described. For ages the sun has made things grow, but solar science can make it grow them better. Reflectors, tilted beds and heat-retaining caps for plants are a reality. Dye which makes for better operation of the solar still can also melt snow faster so farmers can plant earlier in the year!

We have talked about solar furnaces of many types, but perhaps the most unusual is that employing a parabolic mirror of mercury. Since mercury in its normal state is liquid, the problem of forming a parabola of such a material seems insurmountable. The solar scientist simply rotates a vessel of mercury, however, and physical laws cause a depression to form!

The same phenomenon can be observed by simply turning a pan of water, though of course in the application mentioned above accurate control results in precise focus and temperatures. The Japanese have applied this principle to an economical process for forming plastic parabolic reflectors. Liquid plastic is poured onto the surface of the water as it rotates. Surface tension spreads the

Hoffman Electronic Corp.

HERE THE SUN is used to drive the propellers of a model plane. Eight solar battery cells on each wing change sunlight into electric power. While such a small collector surface would not provide sufficient energy to fly an airliner, solar batteries could be used to operate auxiliary equipment such as radios, fans and so forth.

plastic, and it hardens into exactly the shape required! These reflectors are used in solar research.

Modern as this technique seems, it actually dates all the way back to 1900. The method is described in a book appropriately called "A New Century of Inventions." As in every science, ideas appear long before they are able to be put into practice. Da Vinci's parachute is a good example of this, coming long before there were any airplanes. In our time, however, we are seeing the result of all the work done in the past.

The photovoltaic process, which changes sunlight into electricity, has long been known. When you use a light-meter, for instance, you apply this principle. As recently as 1935 experts could not work up much hope for such a method of using solar energy because the selenium cell was a very inefficient converter. "Until something fifty times as good comes along," they said, "we can't consider the sun a power source."

Something similar to this happened in the field of aviation. Authorities flatly stated that five hundred miles an hour was the absolute top for aircraft. Today's planes, traveling nearly three times as fast as sound, prove how wrong this conclusion was. New technology, better fuels and materials change our entire concept. The same thing has happened in the photovoltaic field.

Where once a fraction of one percent was the best a converter could do, silicon discs now turn into electricity eleven percent of the sun's energy striking them. Bell has proved solar batteries can provide part of the power for telephone lines. There will be many other uses.

A dozen photoelectric cells power a toy automobile. There are also model trains, airplanes and even an air-

Hoffman Electronic Corp.

SOLAR BATTERIES run model train. Scientists have suggested the use of such collectors, installed along the railroad right of way to drive full size trains. At present, solar batteries are too expensive for such an application, but research is constantly lowering the price.

craft carrier driven by solar energy. Mechanical magazines give plans for hobbyists to build solar motors and portable radios. In one application, a small square of selenium photocell turned a motor at 400 revolutions a minute, developing one watt of power.

A solar generator in the form of a one-eighth inch square of cadmium sulfide has been made to drive a small motor, producing one third of a volt of electricity. There is a serious drawback at present to using such a motor industrially, because the materials are expensive.

The silicon used in the solar battery originally cost up to $400 a pound, enough to make anything but laboratory experiment out of the question. It is well to remember, however, that aluminum once cost $500 a pound, in a day when a dollar was worth much more than now. Silicon is plentiful, and improved methods have already lowered the price to $180; perhaps further reduction can be expected.

Solar energy makes ice at a laboratory in Tashkent, Russia, and is used in a sewage disposal plant in Tucson, Arizona. One firm is experimenting with paints and wall-papers which "soak up" sunlight during the day, and then give it off at night. No job is too big or too small for the sun!

Artists and craftsmen have long applied powdered enamels to metal by heating to high temperatures, and now some of them work with a solar-fired kiln instead of the usual electric oven. The solar kiln makes use of a plastic Fresnel lens to focus the rays of the sun into a small crucible of firebrick.

Such a kiln can be made for a few dollars and is an interesting project for home or school. Its use is simple,

Mellmann Photo

USING THE SUN to "fire" enameled jewelry, the artist has placed a copper shape, coated with enamel powder, in the brick oven. It glows brightly because the lens at the top of the wooden frame focuses the rays of the sun into it. The "hot spot" is brilliant enough to require the use of welding goggles, and melts and fuses the powder to the metal.

requiring only that the sun be focused properly. Wearing goggles to guard against the glare of the 2,000 degree hot spot, the artisan places the metal to be fired in the crucible. The high temperature flows and bonds the powdered enamel to the copper shape; giving effects not attainable with regular firing methods.

Another use for the kiln is in silver soldering, brazing and other operations. The home craftsman can build such a device for very little, getting excellent results since the solar heat is pure and uncontaminated as has been noted before.

One inventor made a cigarette lighter of dubious merit by simply placing the cigarette through a flashlight reflector. This is perhaps the most trivial use the sun has been put to, in comparison with tasks like operating a huge still as large as an acre in area. But big job or small, the sun is impartial, pouring down its allotted heat on each square foot of earth.

Each day there does seem to be something new under the sun, some fresh application of solar energy. Where once we had a handful of men working separately to harness the sun, now international organizations promote the exchange of ideas and industry backs the research necessary. Many branches of science have a stake in further knowledge of the sun, and they are working together toward that goal.

The big fireball in our sky is a challenge, and the waning woodpile gives an added urgency as man turns to the sun. Ages ago we were born of the sun; today we live in the sun. When we at last reach out into space it may be solar energy that lifts our ships from earth.

CHAPTER SEVEN

Food and Water from the Sun

So far we have talked of the sun as a means of providing power to take the place of fuels that are slowly running out, and as a means of heating homes and possibly cooking our food. But what of the food itself? The same inescapable pressures that make the fuel experts worry about the future are producing a nightmare for those concerned with feeding our fast-growing population. Every day there are 60,000 new mouths to feed, and a shiny new solar stove will be a grim joke if the cupboard is bare!

We commonly think of huge storehouses filled to overflowing, and grumble about surpluses. The truth is that two years of drought, similar to the actual happenings of the 1930's, could wipe out our surpluses of grain and other foods. A member of the Senate Agriculture Committee recently stated that production of food is running only three or four years ahead of population.

The United States is one of few lands having more than enough food. In many places it is already nip and tuck between appetite and food supply. The prospect of tripled population within a hundred years underlines the fact that of all the fuels the most vital is fuel for us humans, our food. Here again we are dependent on the sun, and to the sun we must look for the improved methods of crop-growing needed to keep us alive.

Photosynthesis, the growing process of plants, is still a secret only nature knows. Somehow chlorophyll is able to change water, carbon dioxide and energy into carbohydrates and oxygen. Carbohydrates, of course, are food. So what we eat is literally manna from heaven, beamed to us through ninety-three million miles of space.

The magic of photosynthesis is the most efficient conversion of sunlight into chemical energy and changes about a fourth of the sun's rays into something we can use. However, this ratio is a theoretical one obtainable only under laboratory conditions. Under actual growing conditions some plants are only one tenth of one percent efficient. Sugar beets, the best, turn only two percent into food, and the average is about one-half percent.

Agronomists, crop production experts, are understandably interested in increasing the yield of crops, and much progress is being made. Better strains have been produced and scientific use of soil and fertilizers helps too. More recent techniques include tilting seed beds so that the sun strikes them more squarely, and using reflectors and heat-retaining caps over plants.

One ingenious trick, mentioned earlier, involves applying dye to spring snow to cause faster melting. Hydroponics, the growing of plants in chemical-filled gravel beds, shows promise and experiments have been made with stacking such beds one above the other to make more use of available ground space.

Association For Applied Solar Energy

ON THIS EXPERIMENTAL farm in California, the seed beds have been tilted so that the rays of the sun strike them more directly. The increased heat energy results in higher crop yield. Paper reflectors help use solar energy more efficiently.

All these devices are steps toward increased production, and in laboratory conditions the yield of beets in Holland has indicated the possibility of harvesting up to forty tons of dry plant material per acre per year. This is so much more than the normal crop that it makes exciting news. However it is still in the experimental stages and it will be a long time before such results are common.

Even the most advanced of these techniques requires proper soil, fertilizers and weather conditions. These factors limit considerably the land available for the raising of food for the world, and already we can see the cultivation of practically every square inch of land as in Japan, Okinawa and other places. Because of this another kind of "farming" is being seriously considered to make better use of solar energy so we will not starve in the years to come.

In 1951 the Carnegie Institution in Washington asked the research firm of A. D. Little, Inc. to build a pilot plant for the cultivation of a very unusual crop. Instead of tilled soil, this "farm" consisted of 600 square feet of plastic tubing, with water circulating through it. The "plant" was a tiny organism known as *chlorella pyrenoidosa,* actually algae such as forms on the surface of ponds!

Resembling some strange chemical racetrack atop a factory building, the *chlorella* "farm" was fertilized by carbon dioxide gas and converted sunlight into food in a most satisfactory manner. With figures obtained from three months of operation, the firm was able to predict yields of up to 35 tons of dry algae per acre per year, far more than is achieved with conventional crop-raising methods.

The advantages of such a crop are immediately obvious. In the first place, with a much higher efficiency of conversion, *chlorella* culture yields several times as much food

A. D. Little, Inc.

CHLORELLA, an algae culture, is grown in this rooftop "food factory." Water circulates constantly through the oval shaped plastic container which admits the rays of the sun. Carbon dioxide is added, and the algae multiply rapidly. Unlike other crops, this is a continuous operation. In Japan, chlorella is already used as a food.

per acre of farm. It needs no rich soil, and in fact can make use of wasteland. Freezing and heavy rains would not be the danger they are with ordinary plants.

Chlorella is much higher in protein and fat content, as well as in the amount of vitamins. Besides this advantage, every bit of the yield is usable since there is no stalk, root or leaves to be wasted. Added to these advantages is the fact that there are no weeds!

Finally, *chlorella* requires only water and carbon dioxide for rapid growth, and can be cultured continuously. This makes the farming operation approach the highly efficient level of industrial mass-production. Very little human attention would be required; the farm would be a sort of automation in agriculture.

Although the starving man of the future would be likely to pay any price for food, it is interesting to know the cost of the finished product. Engineers have estimated that in a large-scale operation the dry product should cost about twenty-five cents per pound.

Interestingly the Japanese have reached about the same figure in their experiments. *Chlorella* production in that country has reached the point that actual foods are being made from it. At the twenty-five cent figure, *chlorella* is only twice the cost of soy beans, the cheapest protein available in Japan.

Compared with eggs, *chlorella* costs only half as much per pound, and only one fifth as much as milk. With meats, the new food compares in price to whale meat or squid, both of which are popular in Japan.

If we are to some day enjoy a diet of *chlorella*, its flavor should be of some interest to us. It is here that the product loses out somewhat, since powdered *chlorella* looks like

Association For Applied Solar Energy

THIS RUSSIAN fresh-water solar still produces about 250 gallons of water a day. Such a device permits use of salt water or water that is otherwise undrinkable.

De Wald Photo

SOLAR SCIENTIST Dr. George Löf with drawing of solar fresh water still envisioned by the Department of the Interior. Installed on the seashore, such a system could use the tides themselves instead of pumps. Salt would be a useful by-product.

green tea, and tastes a little like seaweed! Human ingenuity can work miracles in this field too, fortunately, and Japan already has sampled candy, soup, and even ice cream made with *chlorella* powder. The soup is said to be even more tasty than the popular "Miso" soup which is made with fermented soy beans.

The suggestion has been made that *chlorella* be fed to animals since they are not as fussy as humans in these matters. This has been done in the United States, and also in Germany, though of course the high price makes it uneconomical except as an experiment. Chances are good that by the time it becomes necessary to subsist on *chlorella* foods they will be quite palatable, since hunger is surely the best sauce!

Japan is more pressed for food than we are, and she has done more work in the field of *chlorella* culture. In attempts to reduce the cost of the "farm" Japanese scientists have experimented with culture on open ponds or tanks. This has been fairly successful, except that the dust must be removed occasionally, and once in a while a cannibalistic culture invades the *chlorella* and eats up the entire crop in a day or two!

There are difficulties to be worked out, of course. The plastics used thus far are subject to brittleness and cracking after a period of time. *Chlorella* culture uses 4,000 tons of water per ton of crop, compared with 1500 tons of water per ton in conventional farming. But necessity is said to be the mother of invention, and when the need is food itself, we may expect that these details will be overcome. A farmer may someday merely turn a valve and start a pump to harvest his crop.

While it is natural for the hungry man to turn to the

sun for food, we generally think of Sol as making us thirsty and certainly of no help when we crave water to drink. Oddly, the sun will provide fresh water in the middle of the ocean as downed aviators can attest. For this purpose another solar device is used, the solar still.

The whole world is a giant water-purifying system, as a matter of fact. Sun heat evaporates water, principally from the oceans, and this eventually condenses in the form of rain. Salt and other minerals are excluded in this process, a phenomenon that gives us seasoning for our food. Water, food and seasoning all come from the sun.

It seems paradoxical that a man can go thirsty in the middle of an ocean of water, but unlike fish we cannot subsist on salty water. Large ships have distilling plants, but it was not until World War II that solar stills for use at sea came into being in any quantity. The result was a plastic, inflatable and floatable bag which evaporated sea water, then condensed it so that it was drinkable.

A simpler version consists of black toweling, the end of which is dipped in the salt or brackish water. Such stills have been made by students in grammar school science classes, but it is this same principle that furnished up to 6,000 gallons of fresh water in a place appropriately named Las Salinas in the mountains of Chile.

This huge still, covering 51,000 square feet of area, was built 85 years ago and made habitable an area that otherwise would have been wasteland. Interest is high today in like systems for obtaining fresh water in many parts of the world. In Israel, for example, the Solar Radiation Laboratory has worked out a project for supplying fresh water, using solar stills fed from Red Sea salt water.

Even in our own country, some sections are already

Association For Applied Solar Energy

SAILORS USING plastic fresh water stills. These lightweight, inflatable solar stills operate solely on the sun's heat and produce enough fresh water to sustain a downed flyer or shipwrecked sailor.

troubled by water shortage. Arizona is interested in the possibility of using Gulf of California water for irrigation. Of course such water must be de-salinated, and the solar still would be an excellent answer.

So important is the adequate supply of water that the Department of the Interior has let contracts for construction of large solar stills. A still, one acre in size, built of concrete, has been designed. A test model, one tenth this area, is being installed at San Diego, California, converting Pacific Ocean water into something more drinkable.

While the initial cost of a large solar still is high, there is little in such an installation to wear out. It is possible that the tide itself can fill the evaporating chambers so that no power will be required for pumps. A useful by-product in the form of salt makes the idea more interesting. The fact that such a water system operates only during daytime has been pointed out, and also that it will not distill when it is raining. However, at such time it will at least collect rainwater which is already fresh!

In places where there is an abundance of fresh water available at low cost, the solar still is not likely to appeal. There are presently two large fuel-operated distilling plants, one at Kuwait, Arabia, on the Persian Gulf and the other at Curacao, Dutch West Indies. These yield one million and two million gallons of fresh water per day respectively. Cost of construction was approximately six dollars per daily gallon yield.

In comparison with these fuel-operated plants, a solar still would cost between two and four dollars per daily gallon initially, and then deliver water for about one third the cost. Such figures indicate that there is a bright future ahead in the field of water purification by solar energy.

Association For Applied Solar Energy

THIS LARGE SOLAR still, covering an area of 51,000 square feet, was built in Las Salinas, Chile, in 1872. Available water was undrinkable, but the still provided 6,000 gallons of pure water a day with no operating expense. Heat trapped under the glass plates evaporated the brackish water and the vapor condensed on the slanting surface of the glass. Fresh water ran into a trough and then to a storage tank.

CHAPTER
EIGHT

Power from the Sun

It is estimated that the whole world uses approximately five billion horsepower, with the United States accounting for one billion of this total. With the prospect of "running out of gas" in the near future, what will we do about it? There are two obvious choices. Either we must learn to live much as the cave men did, or we must find ways to continue the delivery of power in the huge quantities to which we are accustomed.

It has been suggested that vegetation could be used as fuel when coal, oil and gas are gone. Calculations show that roughly six billion horsepower could be developed in this way, more than enough to drive the wheels of industry. However, we would starve to death as a result.

The idea of growing algae and burning it for fuel has also been advanced, but here again such a product would be far more useful as food. A man having a choice between a meal and an auto trip will probably choose the meal, so fuel experts feel we must look elsewhere for motive power.

Besides the fossil fuels there are a number of so-called "eternal" power sources. One of these is the internal heat of the earth and in rare instances this force has been put to work. Italy has made slight industrial use of this phenomenon, but more frequently we hear of it as the destructive, uncontrollable might of a volcano. Harnessing such energy is a doubtful prospect.

The energy of earth's rotation is tremendous, but it too is difficult to apply to our needs. Gravity, a mysterious power that no one entirely understands, may someday furnish useful work. Several organizations are studying gravity, and "g-engines" have been talked of.

Even the most optimistic researcher does not offer either of these last two energy sources as a practical answer to the power shortage. With all possibilities explored, including fission, which may help industrially while the material lasts, and fusion which seems more suited to engines of war, we turn again to the sun. Solar energy has been rightly called our only "income energy," and on that we must live when the legacy left by nature is gone.

Fortunately the sun is bounteous with her gift of energy, because earth is a prodigal child. Fully a third of the sun's rays are reflected back into space and lost forever. The natural process of rain-making is an expensive one, using 20,000 times as much power as we humans require for all our needs!

Of course we need rain for many reasons, and we do salvage some power from earth's weather factory. With dams and hydroelectric plants we trap about 50 million horsepower all over the world. Potentially there is nearly a thousand times this much water power available to us, though requiring a terrific expenditure for huge installations.

The tides, due in part to the sun, offer another possibility as a power source. Some work has been done to put the rise and fall of the sea to good use, but so far such use is negligible. Here again such projects are costly, depend on water, and require transmission of power if it is to be used anywhere but on the shore.

Windmills for pumping water were a factor in the opening of the American west. They are still a familiar sight in many parts of the world, and a picturesque trademark for Holland. Wind-driven generators provide electricity in areas not reached by power lines, but in the United States these generators, and windmills, develop a total of only about 50,000 horsepower. This is but a tiny fraction of our needs, and even exploited to the utmost the wind could not take care of a significant amount of requirements.

In sharp contrast to the power available from wind and water, which are secondary solar sources, is the vast amount of directly available energy. The engineer can draw on two million *million* horsepower! Since present power consumption is at the rate of only five thousand million horsepower, the sun offers hundreds of times what is necessary. Even more important is the fact that it will continue to do so for ages to come.

We have talked of domestic heating with solar energy, but what of the other three fourths of our power needs? Transportation, industry, and electrical generation must be provided with fuel too, and the solar scientist is interested in this supply that is free for the taking.

Taking it is not quite as simple as it sounds. The sun

Association For Applied Solar Energy

DR. CHARLES ABBOT, standing at right, displays his solar engine. This machine, built in 1936, consists of heat collector and boiler, and develops ½ horsepower. In a special demonstration broadcast, Dr. Abbot provided power to the transmitter by means of solar energy.

beams vast quantities of light to us. In this light are particles of energy, bullet-like "photons" which excite molecules of matter they strike. This results in heat, the basic form of power. For over a century this theory of solar engines has been known, and we have described early experiments in the field of hitching the sun to mechanical devices.

Although Ericsson was aware of the limitations of low-temperature solar heat, many wild-eyed inventors thought a small mirror or lens would produce fantastic amounts of power. The truth is that for each horsepower of energy to be developed approximately fifty square feet of collecting surface is needed. For this reason the simpler the collector and generator can be kept, the more likely the solar engine will be practical.

The flat plate collector has been described in the chapter on house heating, and such a system has been considered for use in a large power plant. A solar power station was designed for El Paso, Texas, consisting of acres of flat plate collectors to produce municipal power. This sunny locality receives the equivalent of almost ten million kilowatt hours per acre, per year.

At the time the research was done, it was estimated that such a plant would produce power for about three times the cost of conventional methods like natural gas. Obviously electricity at three times present rates is not attractive and such a project will not be built this year or next.

Since the El Paso plans were made, however, a development in flat plate collectors called "selective surface" coating has been invented by an Israeli scientist. This coating makes the collector much more efficient, and easily produces steam. It is possible that further improvements

Weiner Photo

WILLIAM RHODES, inventor, describes his Novoid solar engine to Dr. Lee De Forest, called the "Father of Radio." In its way, the engine shown here is perhaps as revolutionary as the radio tube which De Forest invented. Heat from the sun is focused on part of engine being pointed out. Steam is generated, and causes a pulsating hydraulic thrust which can be used to drive mechanical devices.

will make solar electricity competitive with that produced by gas or oil plants. At a recent meeting of the American Institute of Electrical Engineers a paper titled "Solar Energy as a Supplementary Power Source" treated the idea as quite feasible.

A figure has been given of fifty square feet of collector area for each horsepower, but this is for a condensing reflector or lens, which also is able to follow the sun during the day. The flat plate collector which has been considered for an electric power station varies in efficiency from excellent when the sun's rays shine squarely on it, to zero at sunset and sunrise.

It is assumed that in practice not more than fifty horsepower per acre can be realized from a stationary collector system. Based on these figures, it would require an area of about 50,000 square miles to furnish all the power needed by the United States. This sounds discouraging at first but when it is considered that land useful for nothing else would suffice, and also that the solar collector might be in the form of roof surface, the picture brightens.

When other than the flat plate collector is used, the efficiency of the solar engine goes up and is limited only by the second law of thermodynamics. This principle states that heat engines cannot be more than approximately forty-five percent efficient. Allowing normal working loss, the solar engine may still show an efficiency of fifteen or sixteen percent and this is more than most mechanical devices.

Italian solar pumps ranging from one tenth to two horsepower have proved themselves. In this country, solar steam plants have been designed which could be built for about 700 dollars per horsepower. Such devices would

Association For Applied Solar Energy

HENRY CABOT LODGE, JR., U.S. Delegate to the United Nations (*right*), inspects a solar engine made in Italy during a visit to Stanford Research Institute. SRI imported the solar engine, which produces 1½ horsepower, for research on its use for drawing subterranean water to the surface in desert and other wasteland areas.

handle most farm chores now done by electricity or liquid fuel engines, and in remote areas where fuel is expensive because of transportation costs, could compete favorably even today.

Another country which has done much in the solar energy field is Russia. The Krzhizhanovski Institute has built many power plants at Tashkent and worked out the design for a full-scale steam-electric solar plant.

Using a central boiler placed on a tower forty feet high, this installation makes use of an idea dating back to a solar researcher in this country in 1903. Large mirrors are mounted on carriages on a circular track arranged about the boiler room. In this manner the mirrors may follow the sun and keep its rays focused on the boiler to generate steam during the hours of daylight.

Russian scientists estimate such a plant would provide 1200 kilowatts of power, and would take care of the needs of a city of fifteen or twenty thousand people, heating their houses in winter and cooling them in summer.

The Russians foresee the use of small solar engines to provide hot water, home heating, refrigeration, fresh water and so forth; as well as large power installations for irrigation, drainage and industrial use. Quite aptly, they refer to this as "heliopower," and predict a need for it in the very near future if not immediately.

In a previous chapter a French plan for a solar-electric railway in the Sahara was mentioned. A similar idea has been suggested for railroads in this country. On either side of the tracks, solar collectors would gather power to drive trains along at no fuel cost! Such electricity would not have to be sent long distances over power lines as it is now,

Association For Applied Solar Energy

DR. FARRINGTON DANIELS of the United States, visiting New Delhi, India, is shown a solar hot air engine by Dr. M. L. Khanna. This engine, which uses a focusing type collector to furnish driving power, was designed and built by the National Physical Laboratory of India.

and would save losses incurred in that way. For nighttime operation, batteries could store up power during sunlight and release it later on.

There are other solar engines already built which offer possibilities. The hot air engine, invented before Watt's steam engine, is ideally suited to solar energy power source. Ericsson built the first solar "caloric" engine in this country in 1872. Only the cheapness of other fuels kept it from being adopted then. Current interest in solar energy has revived work on the hot air engine, and test models have been built in Holland and in India.

The World Symposium exhibited many other solar engines, including an unusual pinwheel arrangement of glass bulbs filled with liquid. A screen shaded some of the bulbs and unequal heating of the liquid caused the pinwheel to rotate.

Another device, similar only in that it too used the sun for motive power, concentrated the sun's rays on a tiny pulsating hydraulic motor. This unique engine has no moving parts and is adaptable to uses ranging from refrigeration to water pumping.

Although skeptics point to the fact that to compete with a large conventional power plant the solar plant would be unwieldy in size, they forget an important fact in the generation and transmission of power. It is true that a station producing millions of kilowatts is more efficient than one producing only a thousand, but frequently transportation of fuel to the plant, and power from it, eats up a good bit of the output.

Solar energy, on the other hand, is available on the spot almost anywhere. Thus a small village might well have its own electrical generating plant powered by sun energy at

Association For Applied Solar Energy

THIS IS AN artist's conception of the proposed Russian solar steam power plant. Movable mirrors on the circular tracks follow the sun and focus heat on the central boiler. Models of this installation have been tested and it is estimated the full-scale plant would develop 1,200 kilowatts of power, enough to heat a town of 20,000 people.

a cost less than bringing in electricity from a distant hydro-electric plant. Even if the price is higher than coal or oil generated power, we must agree that it will be better than no power at all!

Thus far we have talked only of a rather indirect method of changing sunlight into electricity. The sun's heat is easiest for us to use with only simple materials at hand. Photovoltaic conversion—changing sunlight directly to electricity—will be discussed in the following chapter. We will see what makes the solar battery work, and also learn about another amazing possibility for using the sun's energy.

CHAPTER
NINE

More Power from the Sun

In October, 1955, a Georgia farmer picked up his telephone and made a call. When he had completed his conversation the farmer smiled and said it sounded just fine. For the first time, direct solar energy had powered the telephone lines near Americus, a town south of Atlanta. The Bell Telephone Laboratories experiment was a complete success.

The device that made this possible was a metal panel about a yard square, attached to the telephone poles. A number of tiny, so-called "solar batteries" converted the sunlight into electricity and furnished this power to the telephone line.

In the last chapter we discussed the conversion of the sun's *heat* into power for various mechanical devices including those which would finally turn it into electricity. We come now to the photovoltaic process, which skips the intermediate steps and makes electric power directly. This fascinating principle offers exciting possibilities in the solar energy field.

As we noted before, something new under the sun is rare. The photovoltaic effect was observed by Antoine Becquerel in 1839. This French physicist was followed in his work by a son and grandson, the latter sharing the Nobel Prize in 1903 with the Curies.

While the first Becquerel laid the groundwork for the conversion of sunlight to electricity, it was 1897 before an American was granted a patent for the application of solar heat to "thermo" batteries.

For a long time there have been two similar devices in use; the thermocouple and the photoelectric cell. The thermocouple relies on temperature difference in two metals to cause a tiny current to flow. It has been used for furnace controls, and also by astronomers to measure the light of stars. The photoelectric cell is best known as applied in the photographic exposure meter.

Both of these conversion methods produced such small amounts of current they were of no value except as controls. The development of a practical photovoltaic converter came about in an indirect way.

The earliest radios, the old-fashioned "crystal sets," used simple semi-conductors as detectors. As radios were improved and the vacuum tube invented, the crystal was all but forgotten. Only as higher frequencies were used did the crystal detector come back into its own. Work was begun in the 1930's that led to the production of transistors, small crystal replacements for the vacuum tube that used only a fraction of the power.

The element silicon was used early in this research, but it was found that germanium was better suited for transistors. However, as a by-product of transistor research, scientists discovered what is known as the "p-n junction."

Bell Laboratories

THE LINEMAN is checking on the operation of a solar battery installed on telephone pole in Georgia. In this experiment, Bell Laboratories proved the ability of solar energy to furnish current for telephone service. Used with storage batteries, the unit shown here can provide continuous power. At present the solar battery is too costly to be used in place of regular power, but research is lowering the price so that some day "sun current" may be economical.

By introducing impurities into silicon, researchers created positive and negative materials between which current would flow.

After World War II, work was intensified on the silicon p-n junction and it was noted that it would be ideal for uses such as solar energy conversion. Between theory and realization there were many barriers.

Silicon, coming from sand, is one of the most abundant materials on earth. But ordinary silicon was of no use for a solar battery. Painstakingly scientists had to learn to "grow" single crystals of silicon, and then purify these crystals. Obviously no one man invented the solar battery. It was the result of long cooperation among many men in many fields.

The first successful solar battery was demonstrated in 1954. The device is credited to three inventors; a physicist, a chemist, and an electrical engineer. The battery itself was a strip of arsenic-treated silicon with a thin coating of boron on top.

While the completed battery looks simple, it is difficult to make. The crystals used in its production still cost about $180 per pound, and accuracies of one ten-thousandth of an inch are required in making the battery. Thus far it is too costly to be considered a replacement for standard power methods, but it must be remembered that the device is only a few years old. In time it should become much more economical.

Each square yard of the earth's surface receives approximately one kilowatt of energy. We have discussed solar cookers which deliver about half this, or 500 watts, to the cooking vessel. At first glance this sounds phenomenonally efficient, until we remember the energy is still in the

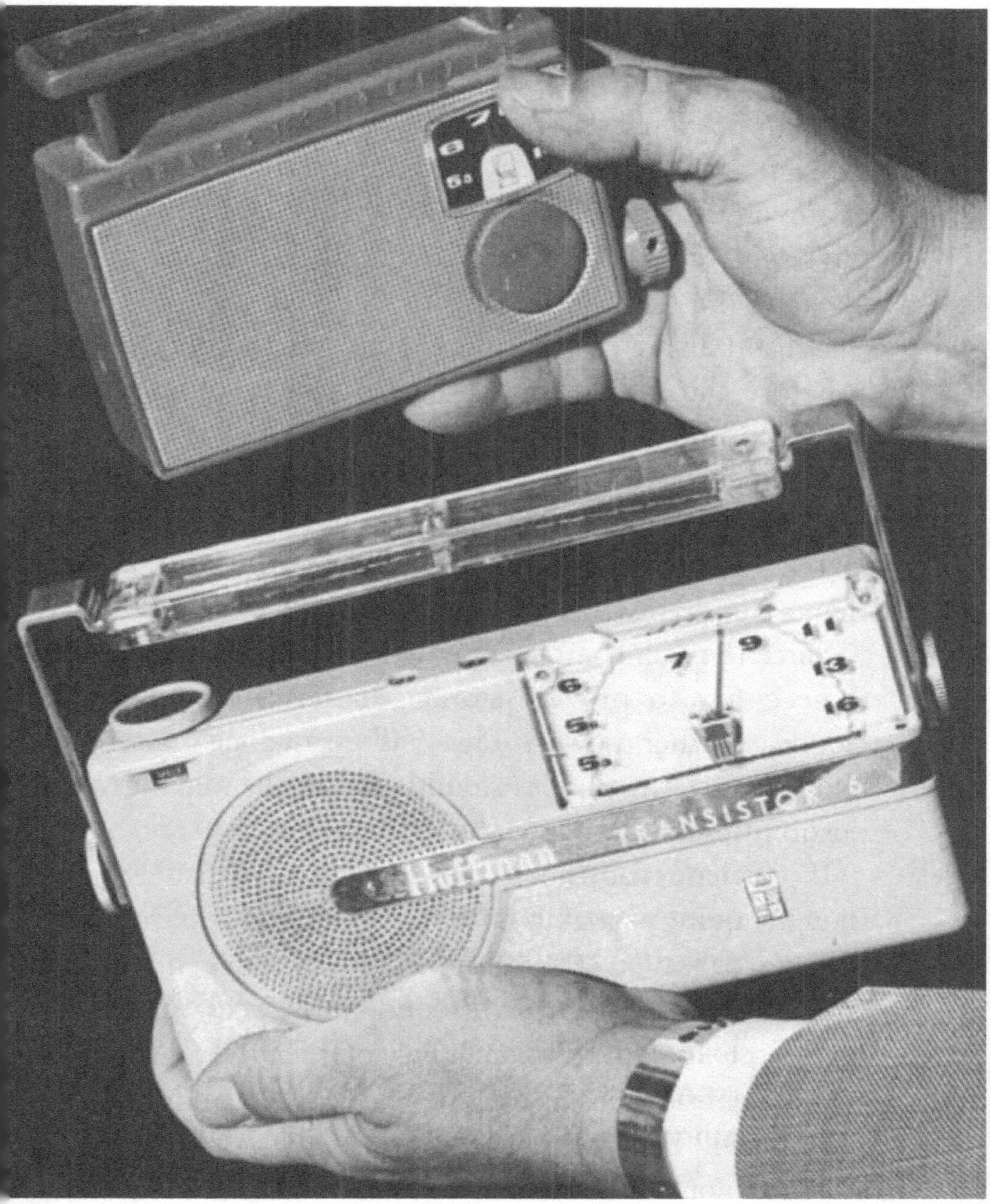

De Wald Photo

SOLAR-POWERED RADIOS. Using transistors, which require very small amounts of power, solar radios operate on current generated in the plastic handle by solar batteries. Surplus current is stored for use when sunlight is not available.

form of heat. Converted into electricity by mechanical means results in only about a hundred watts flowing through power lines.

Scientists believe that a maximum of 22% is possible in photovoltaic conversion. This is because only a fifth of the sun's spectrum gives photons, or "bullets," strong enough to excite the electrons in silicon. The figure of 22%, however, is far above the best of our conventional machines.

Before the solar battery, the best photovoltaic devices delivered only a fraction of a percent of power. The battery demonstrated in 1954 was rated at six percent. In just one year this was nearly doubled, and the batteries used in the Georgia telephone lines converted 11%, or more than a hundred watts per square yard.

Since the solar battery operates only during daylight, it is necessary to provide storage batteries for night and cloudy weather power. This has been done successfully and power is available around the clock, making many possibilities apparent for the solar battery.

If efficiency of the battery were to rise to a figure thought quite probable, installation of 800 square feet of collector area on a rooftop would produce more than enough electricity for lights and appliances. Even at a high cost for installation, the solar battery would be economical since it has no parts to wear out and, in theory at least, will last forever!

While none of us is likely to make such a home installation in the near future, there are ways in which we can put the sun to good use with solar batteries. Firms market radios equipped with solar batteries. For the hobbyist, mechanical magazines have published plans and instructions for solar-powered small motors, and also radios.

De Wald Photo

"BIG BERTHA" of the solar batteries. This panel is made up of 400 cells which convert sunlight directly into electricity. With operating efficiences of better than ten percent, such converters deliver about 100 watts power per square yard of collector surface. This panel, made by Hoffman Electronics, Inc. is designed to track the sun as it moves across the sky.

A short-range radio transmitter operating on a battery the size of a package of cigarettes has been demonstrated, and such units would be useful for campers, rangers, and as emergency equipment for downed fliers.

As an indication of the wide interest in solar batteries, a number of firms now produce solar cells. Research work is being done using germanium and also cadmium sulfide in place of silicon. An electronics firm has installed batteries in model planes, trains and boats to show the possibilities.

Solar batteries atop our garages could charge storage batteries for use in cars when, or even before, gasoline runs out. The car top itself offers an interesting possibility! There are small electric cars available today, and free battery-charging would make them more desirable.

These are interesting speculations, and as such they only scratch the surface of what may someday be. While it is foolish to say that in ten years the solar battery will be our main source of power, it is certain that it will soon do much more than carry the human voice along a telephone line.

To this point we have been discussing the use of devices for changing sunlight into electricity in one step. We come now to another process which makes use of sunlight. Photo-chemical in nature, this process uses the sun's rays to decompose water into its elements—hydrogen and oxygen. These gases both may be burned as fuel, to produce power from the sun!

Where the agronomist is concerned with photosynthesis, and the electrical engineer with the photovoltaic process, it is the chemist who carries solar research ahead in the photo-chemical field. This is the newest phase of solar

Association For Applied Solar Energy

HOW SOLAR BATTERIES may some day furnish electricity to a power line. Using land that would otherwise be wasted, and requiring no tending, the batteries will convert sunlight directly into current instead of the present method that calls for burning fuel to drive generators. A similar suggestion has been made for railroad rights of way, furnishing driving power for electric trains.

energy work, and while it is still definitely in the laboratory stage it is one of the most challenging prospects the solar scientist has.

For years swindlers have held out the bait of making fuel from water by the addition of a magic "power pill." Many gullible people have been separated from their money by such a rosy dream. Since water is the thing we have most of in the world, a means of using it as a fuel would solve all our problems.

An early step in the solution of this age-old dream was the "Hill Reaction," discovery of a British biochemist. It was found that chlorophyll-carrying bodies when illuminated will accept electrons from water. This was a ray of hope in the search for a way to decompose the water into hydrogen and oxygen, but only a faint ray since the reaction was only temporary.

A Russian physical chemist thought of using compounds other than water, from which it was easier to remove electrons. Here again, though, the reaction was brief, and then the compound returned to its original state.

Massachusetts Institute of Technology scientists have been working on the problem for some time, and have devised a method of adding a coloring agent to water for improved decomposition. Thus they have been able to break water down into its component parts.

Ultra-violet light has been used in these experiments, which means only a small part of the sun's spectrum is effective in the process. In spite of this fact it is expected that the process will have an efficiency of thirty to forty percent, using techniques being studied at Stanford Research Institute.

The chemists working on this problem are optimistic in

Hoffman Electronic Corp.

THE FOUR DISCS showing on the deck of this model carrier are "solar batteries." They convert sunlight directly into electricity and drive the motor in the ship. Notice the wake behind the carrier. Solar energy could not power a ship of this size, but might drive auxiliary equipment and also distill drinking water.

spite of the fact that so far only small amounts of gas have been liberated in experiments. They feel there is no good reason for believing the decomposition process cannot become one of economic importance.

There are two distinct possibilities in the solar photochemical field. First is the breakdown of water into hydrogen and oxygen. Second is the generation of direct electric current in the solution itself.

In England two researchers have proved the possibility of recombining hydrogen and oxygen through a porous fuel cell. This results in water, and also the generation of electrical current with as high an efficiency as sixty-five percent!

It is easy to see how simple it would be to join the two processes for the production of electricity from sunlight. Using realistic figures for both steps in the conversion an over-all efficiency of about twenty-five percent might be realized. This, we should remember, is much higher than the most optimistic prediction for the solar battery.

The English scientists, working at Cambridge University, have operated their fuel cells for a thousand hours. Since the cells and the fuel they use are very light their possible importance becomes obvious, particularly in the transportation field. This method also takes care of the problem of storing solar energy very neatly.

While perfecting the decomposition method will by no means be easy, it is certainly not impossible. The cracking of gasoline by fractional distillation, the production of synthetics, and many other industrial processes now taken for granted were at one time seemingly insurmountable hurdles.

By spending billions of dollars, man has unlocked some of the secrets of the atom. Solar scientists are confident that a fraction of that sum applied to research in their field will yield equally fruitful results, with the added advantage that such results contribute to the maintenance of life rather than to its destruction.

CHAPTER
TEN

The Solar Scientist

It has been only a little over ten years since a fantastic, mushrooming cloud turned the attention of the world to a newcomer in the field of science. This modern expert and the tools he works with were created by the spending of huge sums of money, none of which has yet been recovered except militarily. Even when atomic power plants do begin to send electricity through the wires it is possible that this new technology has a limited life. This is because atomic fuel, like coal and oil, can be used just once.

In contrast, the solar scientist dates all the way back to Archimedes, more than two thousand years ago and his future is bright. It is as though the sun has tried for centuries to tell man to look up, that he must in the end turn to the sky for his power. The solar scientist, far from being a novelty, is one of our oldest researchers.

We have talked of Lavoisier, the martyred Frenchman, and of Ericsson in America. Other men, famous for their work in different fields, were attracted to the sun. Langley built an airplane and also solar devices. Professor Goddard, the father of modern rocketry, patented four solar inventions.

Even Pasteur, whose name has carried over in the process he gave to mankind, worked in the solar research field. Either these men were all far ahead of their time, or the rest of the world lagged behind them. There was no need, or so it seemed, for the direct tapping of the sun for power. When the forests were gone, man could burn coal and oil and gas, couldn't he? And brilliant men would dig further into the earth, or the atom, when those fuels were gone.

Today, man has begun to look toward the sky. At the World Symposium in 1955 it was evident that at long last the sun was being taken seriously. Interest in solar energy burst out with explosive force, seemingly triggered by international gatherings of scientists first abroad, and then in the United States. Where men had once worked alone they now had an association, and journals of their own. Where atomic research was conducted in utmost secrecy and produced tensions that gravely upset the world, solar research enjoyed a free exchange of ideas. Sun energy, as we have mentioned, seems ideally suited to peaceful purposes.

With the distinct possibility of need for solar energy as early as 1975, many of the young people now entering the study of science and engineering may well be the "helio-technologists" of tomorrow. No wild claims are being made, and obviously a high-school graduate could not go out and find a job as a solar technician.

That same high-school graduate could go on to college and become a solar scientist, however, because the handful of trained people available now cannot do all the work that is ahead. When the Air Force announced it would build a huge solar furnace high in the mountains near Alamogordo, New Mexico, it invited interested college personnel, *or students,* to apply for appointment to the project.

There will be opportunities in this new field—one with a force at its command as far exceeding the atom as that mighty force outstrips the oxcart! It is a challenge to scientifically inclined young people to begin preparing for a task of vital necessity to our growing world. Physicists, engineers, chemists, thermodynamicists and metallurgists are among those who should find a place in solar technology.

Government and industry alike have a keen interest in solar research. Aiding in the recent World Symposium were organizations including the National Academy of Science, Ford Foundation, United States Air Force, National Science Foundation and UNESCO. Stanford Research Institute and University of Arizona helped prepare the program.

In 1956 a symposium on high-temperature research, featuring solar furnace discussion, was held and drew 500 delegates. Their names were a veritable roster of important industrial firms and government agencies. The Association For Applied Solar Energy held its second symposium in January, 1957.

Besides the aircraft industry, interested in high-temperature research for planes which are flying several times the speed of sound, the melting industry has an eye on the solar furnace as an addition to its foundries.

The power industry realizes the importance of solar energy, first probably as a supplementary source and ultimately one of primary significance. Electrical and mechan-

DR. C. G. ABBOT, pioneer solar researcher, and Research Associate of the Smithsonian Institute, with a small model of his solar boiler. Using a mirror 10 feet in diameter, Dr. Abbot has shown that two horsepower can be developed by the sun. A boiler of his design is now being tested at the University of Arizona.

ical engineers number solar researchers among their members, and papers on the subject have been accepted recently. "Solar Engineer" may be a formal title some day!

The effect of the sun in chemistry and in agriculture has been discussed. One interesting application is a laboratory in Arizona for the testing of paints, fabrics, dyes, wire insulation, and many other materials. Interestingly, one measurement of radiation used in these tests is called a "langley," named for S. P. Langley, the pioneer sun researcher.

The director of this exposure laboratory tests everything from plastic pipe to lawn furniture. His instruments include "Marvin sunshine switches" and "pyrheliometers." Besides this, he of course has a thorough background in solar theory and is a practicing solar technician.

The sun is of vital importance in the field of meteorology too, and the French scientist, Lyot, gave a great boost to weather study when he perfected the coronagraph. This instrument permitted viewing the sun's corona and was a great stride in the search for a sun-weather link.

In this country men of the U. S. Weather Bureau proved the connection between sunspots and rainfall. Scientists in Brazil found the same relationship in that country. Today scientists of the High Altitude Observatory in Colorado continue the work. Meterologists use solar observations in aircraft flight planning, another example of how the sun reaches into many phases of modern life.

Communications, the network that makes modern civilization possible, is affected by the sun too. A major part of the observations made during the International Geophysical Year will be made by solar scientists, and the first satellites will take readings of solar radiations high above

Association For Applied Solar Energy

SCIENCE TEACHER points out exhibit to school children, some of whom may become the solar scientists of tomorrow. It is thought that as early as 1975 fuel shortages will be felt throughout the world, and for this reason uses of solar energy are of great importance.

the atmosphere. Another approach to communications has been mentioned in the use of solar energy to power telephones and radios.

Even the architect is concerned with the sun, as is the heating and refrigerating engineer. The process of distillation opens up another whole solar technology too, from the production of salt and fresh water to the disposal of sewage, using sun energy.

It is difficult to think of a field in which solar research is not important. Let us see what it takes to make a solar scientist, and what is being done in colleges and universities. Merritt Kastens, Assistant Director of Stanford Research Institute and much interested in solar energy, has this to say:

"To produce a creative research man in the educational sense, it is necessary to combine a sound elementary orientation in the high school with seven or eight years of university training, and perhaps as much as five years research apprenticeship before profound and significant contributions can be expected. In view of this situation, students now entering high school may be expected to begin to have a significant impact on our industrial technology not much before 1975. The implication is obvious that the men and women who will be called upon to make the greatest strides in solar technology are those now beginning their training.

"This does not imply in the least that active efforts are not now under way to harness the sun's energy, nor that major strides will not be made in this direction during the next 20 years. In fact, one of the most impressive charac-

Association For Applied Solar Energy

SCIENTISTS FROM ALL over the world attend the World Symposium on Solar Energy, in Phoenix, Arizona. Such gatherings, and the forming of the Association For Applied Solar Energy provide for an exchange of ideas in solar research. Interest is increasing rapidly and great strides are being made in this new scientific field.

teristics of the field of solar technology today is the spectacular rate of increase of research effort expended.

"It is perhaps pertinent to point out that even with the current expansion of research activity in this field, total expenditures to date cannot exceed a few million dollars. It is a little disconcerting to compare the state of the art in solar applications after the expenditure of this modest amount of effort with the technoeconomic status of the peaceful uses of nuclear energy after the spending of literally tens of billions of dollars. It is no doubt fruitless if somewhat intriguing to speculate on what might be the state of development of solar technology if equivalent billions had been expended in its exploration.

"However, it seems fairly obvious to me that by the time the current science and engineering students become contributing members of the technical community, there will be an appreciable demand for skills and knowledge in the application of solar power. Contributions from many disciplines will continue to be required to increase utilization of the sun's power. However, just as we now increasingly recognize the nuclear engineer as a special class of technologist, we will probably some day develop the specialized 'heliotechnologist.' "

Until recently there have been no formal courses taught in schools pertaining to solar energy specifically. However in 1950 Massachusetts Institute of Technology offered a session for engineers in the field of solar house-heating. In the summer of 1956 the University of California at Los Angeles gave a course titled, "Utilization of Solar Energy" carrying two hours of credit.

More such courses may be expected with the eventual possibility, as Mr. Kastens pointed out, of formalizing solar

Association For Applied Solar Energy

STUDENT AT LEFT watches solar scientist adjust solar furnace atop Science Building of Arizona State College at Tempe. This furnace, the first in the United States to be equipped with the sun-tracking heliostat at the left, does government research in the high temperature field.

studies. Meanwhile many schools have active programs now in solar research that give opportunity for interested students.

Arizona State College at Tempe, for example, recently installed the first heliostat-equipped solar furnace in this country. This is being used to do research under contract to the government. A 24-year-old physics major had the distinction of being the school's first solar energy student. Along with him, two high school teachers studied and operated the furnace, and will pass on their knowledge to their own classes.

University of Wisconsin has a Rockefeller Grant and works principally with solar cookers, pumps, coolers, and radiation measurements. New York University has a solar program too, including experimental stoves and solar stills.

MIT studies radiation with its furnace, and another project is photo-chemical research which has been described in more detail in the last chapter. Fordham has a solar furnace, and plans a much larger one. The University of Florida includes solar water heater research among its projects.

Stanford Research Institute has a threefold program including photosynthesis, flat plate collectors, and high-temperature research. These are not all of the institutions with active programs, of course, but give an idea of the scope of solar research.

Men in three dozen countries are looking to the sun with an eye to putting it to work for them. Unlike buried fuels which are plentiful in one land and scarce in others, sunshine is generally impartial with its blessings. The so-called "Solar Belt," from the Equator out beyond the

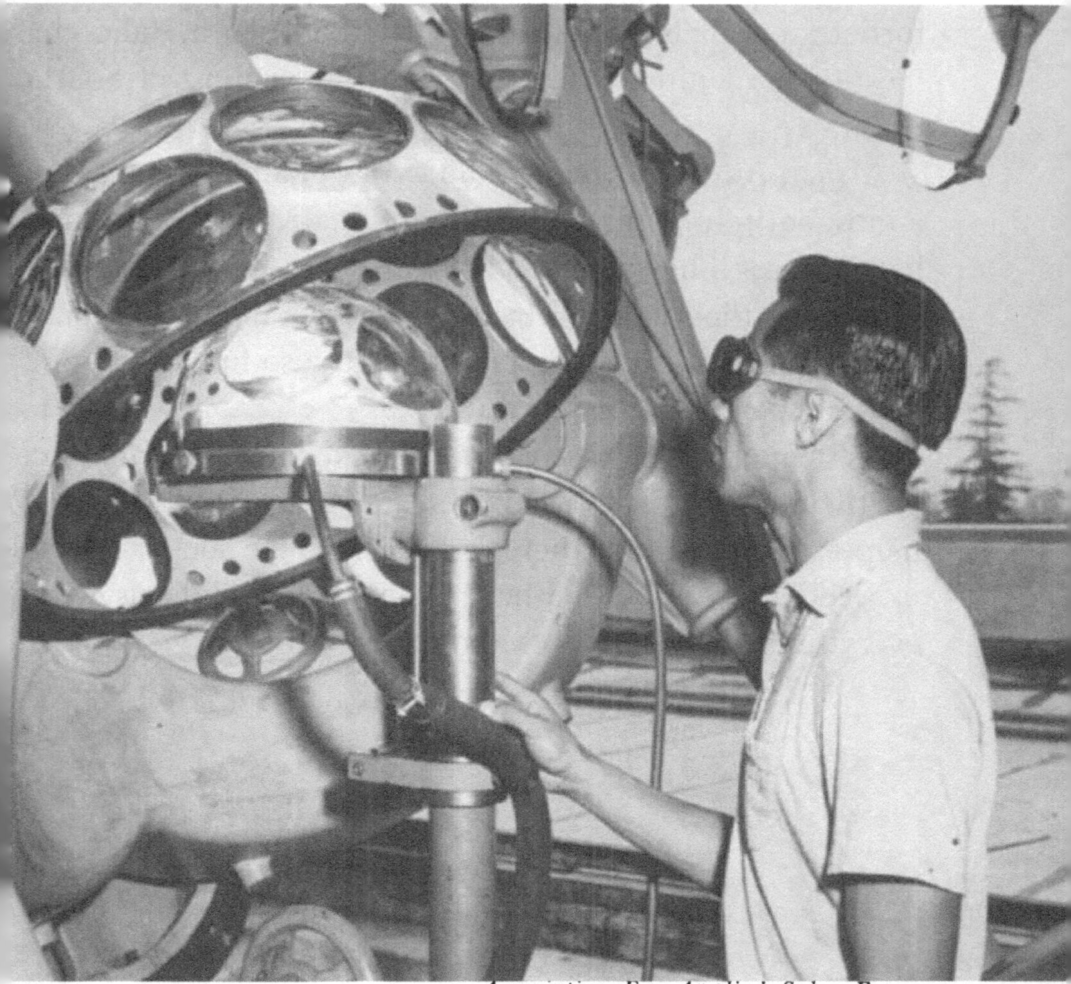

Association For Applied Solar Energy

YOUNG SCIENTISTS like this one operating the large solar furnace at California Institute of Technology are preparing to carry on the research and industrial tasks of the solar energy field. Government and business alike have a keen interest in new developments and offer financial and other aid to the program.

fortieth parallels of latitude north and south, takes in most of the United States. It also includes most of South America.

Africa, Australia, India, China and Spain are countries almost entirely in the solar belt, and many other lands are partially so. The same sun that operates Trombe's huge furnace in the south of France cooks rice in Japan. Solar energy that pumps water in Italy will also fire jewelry in Phoenix, Arizona.

As scientists from many countries exchange information on their latest solar discoveries and inventions they are helping to prove that there is a place in the sun for all of us. And the sun, as we have seen, can provide us with everything we need to live and to enjoy life.

CHAPTER
ELEVEN

Solar-Powered Space Ships

In 1656, more than three hundred years ago, Cyrano de Bergerac described a solar-powered space ship in his "Comical Histories of the States and Empires of the Sun and Moon." At a time when the Pilgrims were just getting a foothold in America, a hundred years before Watt began fiddling with his steam engine, and two hundred years ahead of Jules Verne, who is generally thought of as the first science-fiction writer, Cyrano blueprinted space travel.

The longnosed Frenchman, famous for his writing and his swordplay, was immortalized by Rostand at the turn of the last century as a romantic hero. Cyrano's imaginative foresight was amazing, since he pictures his space craft as a double-walled, hollow cylinder with lenses fitted into the outer surface. These lenses focused the sun's rays to heat air inside, resulting in motive power for the space ship.

Technically the author is guilty of two errors, though we may blame this more on a joking attitude than ignorance. Cyrano must have known that there is no atmosphere in space, and this alone would have grounded his machine for extra-terrestrial flight. Also he described the heated air as rushing out the *front* of the cylinder and thus tugging the craft along with it. Either Cyrano was aware of a principle not yet discovered by scientists today, or his space ship would have flown backwards!

Even with these errors, we must still admire the early writer for his prediction, since as we enter the era of space flight considerable attention is being given to solar power. One example we should shortly see is the use of solar batteries in the earth satellites to be launched.

It has been pointed out that the power available in space will be not the thousand watts per yard we have considered on earth, but nearly half again as much. Even a hundred watts of power is sufficient for the larger, instrumented satellites and would provide energy for radio transmitters, cameras, and other equipment. Since such a satellite will spend much of its time in shadow, however, the solar battery will have to be used together with storage cells to provide continuous power. Another qualification is the possibility of deterioration of the solar batteries by cosmic ray bombardment.

Also speculated on is the possibility of using the fuel cell mentioned earlier, in space travel. Nuclear power is being considered for space flight, but the problems of shielding such a dangerous engine are tremendous. If the nuclear reactor were put in a structure separate from the crew's quarters and fuel cell, the reactor could decompose water into oxygen and hydrogen which would be piped to the space ship proper. There the crew would re-combine them to produce power to drive the entire assembly.

American rocket scientist Krafft A. Ehricke has proposed what comes closer to being an actual solar-powered space ship. Although the picture of this ship does not resemble Cyrano's fanciful creation, there is a technical similarity. The modern design uses a spherical reflector instead of lenses, and hydrogen is burned and ejected rearward. The basic idea is the same—solar heat causing the emission of a driving gas.

The solar-powered space ship was presented in a paper read to the American Rocket Society. It notes that the chemical rocket is hard to beat, but points out that the enormous cost of supplying a space ship with fuel from earth is a serious restriction. Fuel scarcity may well be a more restricting factor than cost, which makes solar energy more attractive as time goes on.

The designer considers nuclear power, mentioning the handling difficulties associated with an atomic pile, and also suggests the possibility of refueling on other planets to lessen the fuel load carried from earth. He believes that the first space travel will be done with chemical fuels, but that ultimately we may use the sun to drive our space craft.

The proposed method, while not 100% sun powered, uses solar energy to replace more than 20 tons of liquid oxygen. Still requiring a quantity of liquid hydrogen, the planned ship would weigh 16,000 pounds and carry a crew of two in a central gondola. The weight of the space ship is thus about that of a fighter plane.

Two huge reflectors, each more than 100 feet in diameter are required to generate the heat for the hydrogen motor, and to construct an assembly of this size, keeping the weight within reason, would be nearly an impossibility. While a parabolic reflector is optically perfect and delivers the highest temperatures this would actually be less practical than a spherical reflector. The spherical shape gives a lower temperature over greater area at the focal point.

Another disadvantage of the parabolic reflector is the unwieldiness of a "dish" the size needed in the space ship. The designer therefore cleverly proposes a complete, balloon-like sphere of plastic. Produced with a thickness of only one thousandth of an inch, the clear plastic transmits 90% of the light striking it. Half the large inflated sphere would be metal-sprayed to form the reflector mirror which concentrates heat on the heaters for the liquid hydrogen.

Filled with hydrogen or helium at low pressure, the huge bag would be cooled as a small amount of gas "diffuses" through it. It is not thought that cosmic dust would make holes large enough to be dangerous.

The smaller spheres at the surface of the clear half of the bag generate electrical energy for pumps and standby power to be used when traveling in the shadow of a planet. The entire assembly can be pivoted to face the sun no matter what direction the space ship is moving.

There are of course many problems involved in such a space ship. The danger of meteor damage is a possibility. Rate of acceleration would be very low, and maneuverability poor. It is felt the design presented here would be useful only for lunar trips because of the limited amount of fuel.

Along with the disadvantages there are advantages however. The fuel, hydrogen, will be readily available in the post-fossil-fuel age. The idea of using the craft as a balloon to great altitudes is intriguing, but the designer suggests ferrying it disassembled to a satellite. Space travel would begin from there, and launching would be much

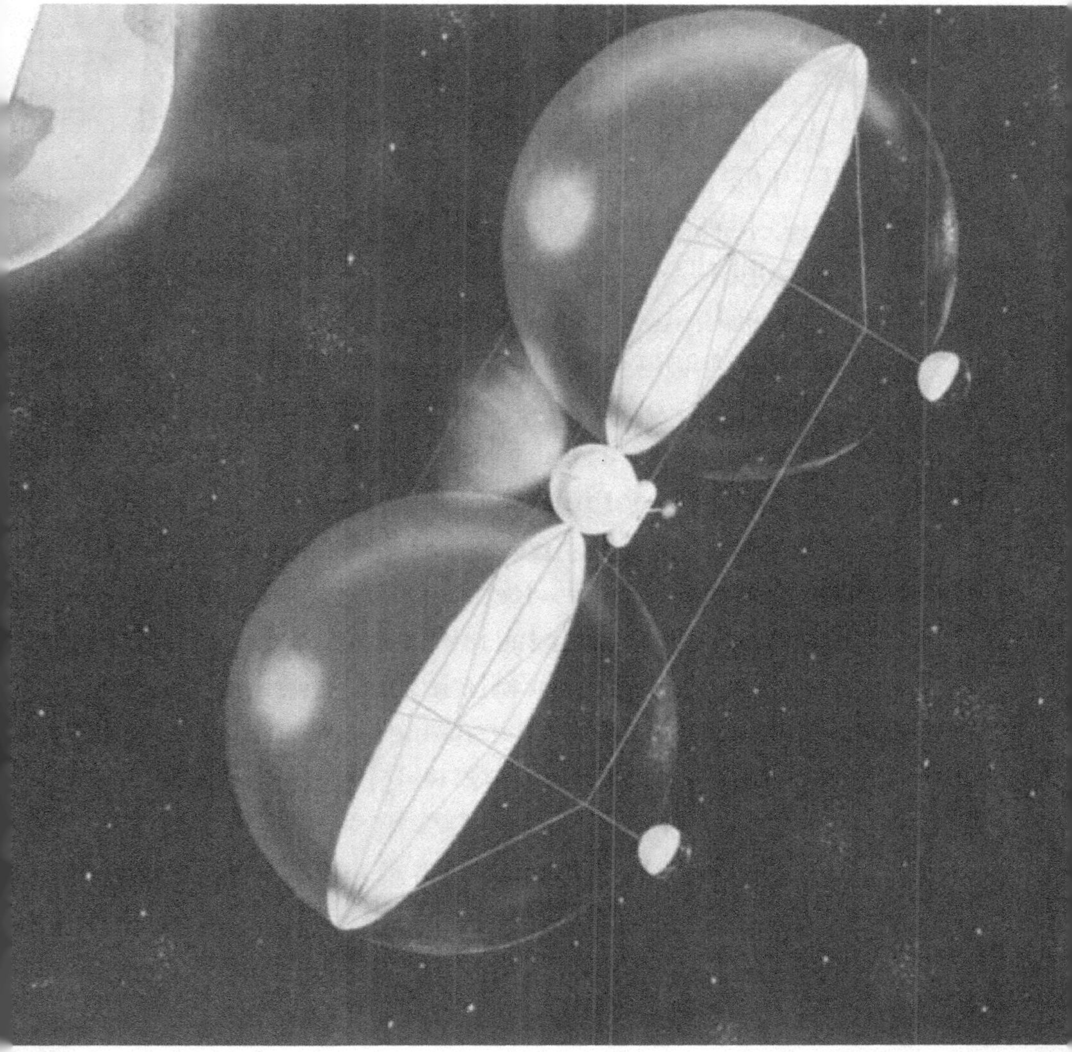

Krafft A. Ehricke

IN THIS PROPOSED space ship solar energy takes the place of many tons of fuel. The two ballon-like plastic spheres are silvered on half their surface to form huge reflectors. Liquid hydrogen is heated by this "solar furnace" and drives the rocket ship. The two small reflectors provide power for auxiliary equipment, pump motors, lights, radio, and so forth. Such a space ship would be very light. Its designer feels it is best suited for moon trips.

safer than conventional high-acceleration methods. For these reasons, *astronauts* may leave earth in a craft resembling the balloons of their pioneering grandfathers, the *aeronauts*.

The proposed space ship uses what might be called a conventional driving motor; that is, it relies on the thrust of a burning fuel. There is another kind of thrust that the space ship designer has thought of, called "photon propulsion." A stream of electrostatic energy emitted from the rear of the craft creates forward movement in this system which is purely theoretical so far.

Dr. Ernest Stuhlinger, who is head of guided missile research at the Army's Redstone Arsenal, has designed such a space ship which he feels could be used for longer voyages through space. While it too uses stored fuel, solar energy is used to heat this fuel and create the emission, resulting in an even greater weight-saving than in the previously mentioned craft.

It is assumed that a large, interplanetary space ship will begin its trip from a satellite of earth, and "land" on a like satellite at its destination. Fuel for the journey would have to be carried from earth to the satellite in rockets, burning up 170 pounds of fuel for each pound deposited at the satellite. Thus cost and scarcity of fuel could make long space voyages prohibitive.

These difficulties led to the idea of an electrically propelled craft. Its engines would use the material cesium or rubidium, which when striking an incandescent platinum surface cause a stream of ions to flow. This ion stream would be accelerated by a negatively charged electrode, making a sort of electronic jet or blast. An example of photon propulsion in our immediate experience is the

Official U. S. Navy Photograph

DR. J. P. HAGEN, Director of the Vanguard Project, and models of the first earth satellite. First use of solar energy in space will probably consist of solar batteries installed in such a satellite to provide power for radio, television camera, and other equipment.

cathode ray tube of television sets, which fires a stream of ions at the picture screen.

Enough such emission to drive a space ship, even at a very slow acceleration, will require large amounts of power. For this purpose the designer would call on the sun. The plans show forty reflectors, each fifty feet in diameter, arranged in roughly the shape of butterfly wings on each side of the crew's section. Such a design results in an unwieldy shape suitable only for travel in free space away from strong gravitational stresses.

Each of the reflectors concentrates its rays on a boiler at the focal point, and the boiler operates a turbo-generator producing 200 kilowatts of power. A large number of relatively small reflectors are used to guard against crippling loss of power because of meteor hits. There is another reason, too, for the smaller reflectors. It is far simpler to build a fifty foot mirror than one hundreds of feet across.

The turbo-generators supply the 7500 kilowatts needed to create the ion drive, with enough left over to power auxiliary equipment such as pumps, radios and so forth. Streams of ions are emitted from 1,000 propulsion chambers arranged in a honeycomb shape. The entire collector assembly is designed to rotate to face the sun and the individual reflectors turn also to aid in circulation of the liquid used in the generator boilers.

Since so many reflectors are needed, along with the piping for liquids and transmission of power to the center control section of the space ship, there are inevitable losses. Besides this the design is not particularly maneuverable. If the solar battery, with its direct conversion of sunlight to electricity, can be raised considerably in efficiency it would aid in making a more compact craft.

Although photon propulsion has many drawbacks, including a very low rate of acceleration which would make necessary the use of power for half the trip for forward motion and the remainder for braking, it has advantages too. Proof of this lies in a hypothetical voyage to Mars.

Carrying a crew of ten, and a "payload" of 50 tons, the electrostatic space ship would make the trip with only 250 tons of fuel. A conventional rocket ship would require 1100 tons. Even more attractive is a longer flight, and one requiring a time of two years would use up only 275 tons as compared with 7500 tons of "old-fashioned" rocket fuel! The longer the flight, the better for the solar space ship.

Since all this fuel must be ferried up from earth to the satellite at a cost of 170-to-1, the chemical rocket space ship would need 1,275,000 tons of fuel! The electrostatic craft would use 46,750 tons, which is bad enough. It may ultimately be necessary to rely entirely on solar energy for flights deep into space.

It was three hundred years from the fictional space ship of Cyrano to the serious proposals of today's astronautic scientists. How much additional time will be needed for making theory into accomplished fact we cannot say with any degree of certainty. There are many who scoff and sneer at the mention of space travel, unable to understand why men should want to venture off the earth. But the adventurous spirit is strong, men have always had an eye on the stars.

What began with Icarus and Daedalus carried on with the Wrights and Lindbergh. Goddard and Oberth blazed trails in rocketry, trails that looked more inviting as the last frontiers *on* earth fell. Now there are scientists who can put what was once a dream on paper. Tomorrow the blueprint may become reality and men will reach the moon. Beyond that are the planets, and the sun shines on them too.

CHAPTER
TWELVE

The Sun Tomorrow

Up to now we have been discussing the history of solar energy and its status today. Important as these phases are they are the faintest shadow of what is surely to come. If the sun is vital to us now it will be more so tomorrow and each succeeding tomorrow. Where the past and its heritage are the prime factors with the fossil fuels, the future and its promise are the aspects of solar energy we are most concerned with. In this last chapter we will discuss the fruits of research being done today.

At the World Solar Symposium, a speaker told the gathered scientists something he heard from an Indian in Canada:

"Sooner or later people will have to learn to live in a way my ancestors did, in a way that does not use up the irreplaceable resources of the country."

This statement, coming from a man close to nature, has the force of prophecy. Solar energy will supplant conventional sources first in remote places, where fuel is scarce or expensive because of transportation costs. India is such a place, as are regions of Mexico, ripe for solar devices to take the place of wood and animal power.

Providentially there is ample sunshine in many of these depressed lands. Here solar stoves and pumps, as well as other machines, will first have practical use. But slowly such devices will find their way back to more civilized countries as other fuel supplies begin to disappear.

At the time Edison invented the electric light no one accurately predicted the future of that glowing bulb. Neither was the fantastic radio and television industry foretold when De Forest gave the world the audion tube in 1907. Similarly it is impossible to say just what life will be like a hundred years from now, but we can make some fairly "educated" guesses. The picture we see is fascinating.

It is the year 2,050. We are living in the United States, population over 400 million. Surprisingly, cities are no larger than they were in 1950, although many of these large cities are now in the southwest. Most people live in smaller communities, spread over a wider area than in the past. It is in such a place, a town we will call Solarville, that we find ourselves. Let's spend a typical day there.

The electric alarm wakes us, and as we dress we notice the sun is just peeking over the horizon to the east. It is midwinter, and the temperature outside is a chilly 35 degrees. The house is comfortably warm, though, and we hear the faint movement of air through the wall registers. The air conditioner stores heat chemically, and even two consecutive cloudy days last week were no bother.

Lights come on in the bathroom at the flick of a switch

and the electric razor hums busily. Glancing out the window we look across the ten-foot stretch of water, seeing trees and other houses similar to ours, but no power poles or unsightly wires.

Atop our house, serving the double function of roof and power supply, is the slanted collector area. Its angle was carefully planned to be most effective, and the house itself floats in the circular pool of water, turning slowly during the day to follow the sun.

This is a much nicer home than the one we moved from last year. The old house was built on solid ground, with a stationary collector so that there were times when the storage batteries ran dangerously low and we had to use power sparingly. This new model is one of the finest available and we are very proud of it as we go into the kitchen for breakfast.

Bacon and eggs smell delicious cooking on the electric range, and the toast is kept warm in the solar oven alongside. As we sit down we listen to the radio describing plans for the first space trip to the planet Mercury. It will not be a simple trip like the scheduled runs to Mars or Venus, and the space ship is powered by the newest solar drive.

As we get up we cast a shadow across the radio and its volume fades until we pass. Smiling, we think about the kidding we take for being attached to the ancient portable that grandfather gave us so many years ago, but it still plays well. Of course the table model radio and the television set are plugged into the house current and are not subject to fading.

The house has turned a little since we got up, but we are used to that by now and it would seem odd if it didn't. Actually it makes for more variety in the view out the circular, insulated-glass walls. Returning a pitcher of juice to the refrigerator, we carry dishes to the sink and stack them in the dishwasher.

The washing machine and dryer are in the kitchen too, using power and hot water supplied with solar energy. The water is sterilized of course, since we pump our own out here in the suburbs. As we slip on a jacket, the air conditioner goes off. With the sun up now, enough heat is coming in through the glass walls to warm the house nicely. The roof will store heat for reserve, and then make power for a good part of the day.

Stepping out into the brisk morning air we circle the house until we reach the access walk. As we go across, the water in the pool doesn't look as inviting as it did in the summer, although if we did want to swim in winter we could warm the water quite easily. Besides its function as pool and turntable for the house, the water serves as heat storage tank and as part of the summer air conditioning cycle.

Passing through the low hedge surrounding the pool, we walk to the garage to get the car. Our garage is stationary, its fixed collector roof providing ample power to recharge the car's batteries, and for the little hobby work we do with power tools. Some friends with several cars and a big shop have automatic tracking collectors even on their garages, however.

Behind the garage the solar pump is busy, and we remember to go wind the clockwork motor that will turn its collector several times during the day. When we first moved away from city-provided water we had some doubts about maintaining a pump and the other hardships of country living, but now we smile when we remember. The

Association For Applied Solar Energy

SOMETIME IN THE future surburban homes may look like this. Floating in a pool of water, the entire glass-walled structure rotates so that the collector mounted on top of the roof always faces the sun. Power for lighting and appliances, as well as complete air conditioning, is provided by the sun. The pool serves for swimming and also heat storage during the winter.

saving on utilities is more than enough to make up for the cost of the car we commute with.

Lifting the garage door, we go inside. There is a faint hum from the battery charger and that reminds us to check the condition of the batteries in the car. The needle shows what should be enough power remaining for the trip to work, but to be on the safe side we lift the hood. It takes only a few minutes to lift out the batteries and replace them with freshly charged ones from the rack. Grumbling at this slight inconvenience, we promise to have the alignment of the car's wheels checked.

It is a two mile ride to the business district where we work and the trip takes only a few minutes. The car is small and light in weight, made principally from magnesium. This metal is now one of the cheapest, being a by-product of the huge distilling plants that produce fresh water along the coast.

The traffic lane on the outside is used by interurban buses, electrically-driven and taking their power from the collector strip alongside the highway. This same source, municipally maintained, provides electricity for street and traffic lights as well as emergency charging facilities for motorists.

As we pass the airfield, a transport plane takes off and heads east towards the sun. Powerful rocket motors, fueled with liquid oxygen, or "lox" as it is called, remind us of our job at the fuel plant. With a degree in solar technology, we found a position as engineer with Allied Decomp, where sea water, piped five hundred miles from the Gulf, is broken down into the gases of oxygen and hydrogen.

Since fuel can be made nearly anywhere now, its transportation is a thing of the past. Thus sea water flows

Association For Applied Solar Energy

THIS POWER STATION, built on the outskirts of a city of tomorrow, uses solar energy to convert water intc the gases, hydrogen and oxygen. The gases are then burned as fuel, or allowed to recombine in what are called "Bacon cells" which produce electricity. If the city lies on the coast, it may use sea water, converting a part of it to fresh water for drinking, and using the salt by-product that is formed.

through the pipes that once carried fuel oil and natural gas. Each city of any size makes its own gas fuels, along with numerous by-products. Airplanes require this type of power and some heating is done by gas during prolonged bad weather. Then too, hospitals and many industries use oxygen and hydrogen. Our job with Allied is an important one and we are proud of the work we do.

Parking under the long, shed-like shelter, we pause long enough to attach charging wires to the car's batteries. The shelter is also a solar collector and furnishes electricity for all the cars parked under it. By the time we come from the plant when the shift is over the batteries will be completely recharged. Some places charge for this convenience, but Allied offers it free to employees.

The day goes well, except for some valve trouble in the west wing of the tank building. By noon we have located the cause, re-routed the flow of liquid and made a replacement of the corroded part. The salt content of the water is near the safe maximum and we make a report to that effect. Then we go to lunch.

Our meal is excellent, the meat and fresh vegetables matched ounce for ounce with algae supplement. *Chlorella* processers are so skillful now that it is hard to tell where the cooks have used the various powders, gelatins and liquids. Times have changed a lot from when the government first announced that land farms could no longer feed the country! In those early days *chlorella* all seemed to have a fishy odor in spite of what a cook could do to doctor it.

The afternoon passes quickly, and as the air outside cools, blowers route surplus warmth from the sunny side of the plant to the shaded side. Once it would have seemed

Association For Applied Solar Energy

FOR INLAND CITIES where surplus water is not available, a different kind of solar power plant may come into use. Inexpensive flat plate collectors of ordinary glass trap the sun's heat in this picture. A steam-generating station provides electricity. Notice the power lines leading into the distance.

odd to have to cool a building and warm it at the same time, but this has been standard practice for some time.

Before leaving the plant, we stop at the fountain for a drink of cold water. Water is no longer the distressing problem in the southwest it was a hundred years ago and it is hard to believe people once were driven out of dry parts of the country! Today sea water is piped hundreds of miles to make the desert literally bloom again. Instead of dropping as it did for so long, the water table is rising so that outlying areas now pump their own water as we do for our house.

Back at the car we notice that the charging cables have automatically disconnected and reeled up into the overhead. The needle on the instrument panel tells us the batteries are fully charged so we drive away from the parking shed. There are quite a few cars on the road but we compare it with stories read of the old days when everyone worked in a huge city. The thought makes us smile sympathetically. Solar living has brought many bonuses.

By the time we shop a bit and then return home, the sun is fairly low in the west and the air is chilly. Shivering slightly we put the car in the garage, duck around back to check the pump, and then hurry for the house. It has turned almost completely around from the way it was in the morning, the gleaming collector on the roof catching the last rays from the sun.

We cross the access walk and the kitchen door is right in front of us. Inside, the air is warm and the drawn drapes create a cozy effect. The smell of dinner cooking wrinkles our noses with pleasure, and we hear the voice of a newscaster on television. We glance at the screen and see a picture of the new space ship, almost ready to leave on its

Association For Applied Solar Energy

SOLAR ENERGY can convert salt water, or water otherwise undrinkable, into fresh water and make areas once useless available for raising stock. It would also be possible to grow algae by the open pond method and use it for feed.

epoch-making journey to Mercury. Wistfully we think that this is the threshold of a new era and wonder what it will be like a hundred years from now!

Our day in Solarville was fanciful, of course. What life will be like in 2,050 we cannot say for sure, but the things we have described are all based on known fact. When the Wright brothers first flew, not even the wildest fanatic would have dared predict things we now take as commonplace—jet airliners that travel five hundred miles an hour, and experimental planes capable of three times that speed.

The steam engine and other inventions once caused a revolution in our lives, and kept the dire predictions of Malthus from becoming grim facts. Solar energy may not revolutionize things, but it will change them. We will live to see some of those changes; one day other people may live *because* of those changes.

Experiments with Solar Energy

Other books by D. S. Halacy, Jr.

COMING AGE OF SOLAR ENERGY

COMPUTERS: THE MACHINES WE THINK WITH

COPTER COWBOY

DUSTER PILOT

FABULOUS FIREBALL

HIGH CHALLENGE

NINE ROADS TO TOMORROW

THE ROBOTS ARE HERE!

ROCKET RESCUE

WHALE SPOTTERS

STAR FOR A COMPASS

SURFER!

RIPCORD

CYBORG: EVOLUTION OF THE SUPERMAN

BIONICS: THE SCIENCE OF "LIVING" MACHINES

RETURN FROM LUNA

Experiments with Solar Energy

D. S. Halacy, Jr.

Illustrated with photographs and line drawings

W · W · NORTON & CO · INC · NEW YORK

First Edition
Library of Congress Catalog Card No. 69-18892

Published simultaneously in Canada by
George J. McLeod Limited, Toronto
Printed in the United States of America
1 2 3 4 5 6 7 8 9 0

Contents

Experiments with Solar Energy

1

Meet the Sun

It's no secret that the sun provides the energy that keeps us alive. Solar heat keeps us warm, grows our food, and makes our weather; without sunlight we would be blind. But beyond these most important basic purposes, solar energy has additional potential we will one day harness. There are many simple applications that demonstrate this potential, and this book is your invitation to the fun of putting solar energy to work in a variety of useful ways. In the chapters that follow there are plans for solar cookers, a radio, a water heater, a furnace, a model airplane that flies on sunshine, and other projects that will give ample reward for the time spent on them. Before we begin to assemble a solar oven or a radio powered by a sun battery, let's take a brief look at solar energy to see why it is so fascinating and challenging.

The use of silicon cells, or some other type of "solar battery" to convert sunshine into electricity, is an ex-

Solar batteries power satellites that detect nuclear blasts.
NASA

ample of the great strides being made in the solar-energy field. Until a few years ago the sun battery wasn't much more than a hopeful gleam in the eyes of research scientists. Now the batteries exist. Today radio receivers are picking up the steady signals generated by silicon cells carried on spacecraft. Meanwhile, back on earth, transistor radios powered by sunlight have been introduced. Of course these radios will operate in artificial light too, and some of them even store up sun energy during the day for evening use.

When we understand the amazing accomplishments of the sun battery in the few years it has been with us, it is easy to understand why it is the best-known use of solar energy. Able to tap directly the vast reservoir of "free" power the sun beams at us, the sun battery is assured of an important role in man's conquest of space.

Building and operating a solar radio will give you only a glimpse of the many and varied possibilities of the solar cell. For example, electricity is a convenient method not only of using solar power but of storing it too. Sun batteries thus offer a practical twenty-four-hour source of energy when coupled with proper storage batteries. The maximum efficiency of commercial solar cells now available and in use is about 12 percent. In other words the battery is able to convert 12 percent of the solar energy beamed upon it into electric energy. Laboratory results have run as high as 14 percent; radiation physics indicates a theoretical maximum of 22 percent. Even at the 12 percent figure, a roof of solar batteries could easily power the most "electrified" home with only a week of clear weather a month!

The drawback, of course, is the present price of solar cells. The rooftop installation we talked about for electrifying a home would involve an outlay of many thousands of dollars, prohibitive except for millionaire solar enthusiasts. However, since the silicon used in the cells is one of the most plentiful elements on earth, the price may be drastically reduced in the future. In the meantime, don't despair. The cells we'll need for our solar projects cost only a few dollars and will last for many years, with proper care.

Early techniques for using the sun to drive engines were crude by our standards, but pioneer solar engi-

Solar-powered hot-air engine built by John Ericsson a century ago. SOLAR ENERGY SOCIETY

Solar exhibit at Casablanca. Shown are projects similar to those in this book.
SOLAR ENERGY SOCIETY

neers like John Ericsson built hot-air engines almost a hundred years ago. The output of these primitive machines was low, but the challenge of "free" energy was strong. Each square yard of the earth's surface receives about 1,000 watts of power during sunlight hours. Some simple reflectors convert about half of this to useful heat, and when we consider that one horsepower is roughly 700 watts, the possibilities are exciting.

Lake Mead, the giant reservoir behind Boulder Dam, receives more sun energy on its surface than its waters develop as they cascade down to the huge hydroelectric generators! At present this solar energy is wasted, but when we learn how to make use of such sun power we will not need to worry about running out of fuel.

Aside from laboratory test engines, working solar pumps are being produced. Huge sun-heated steam plants have been proposed. What will finally be the most economical solar-power plant may depend on the findings of a new generation of scientists. Whatever form the plant takes, it will use the same energy source as our simple solar-battery-operated electric motor. A motor that will fly a model airplane on sunbeams!

Although the solar battery and other sun-*power* devices are used most often presently, they represent only one of several methods by which we can convert sunlight into usable energy. High in the Pyrenees Mountains of southern France is the Mont Louis solar furnace. Constructed just after World War II, this huge installation is 35 feet in diameter and is used in high-temperature research. It develops temperatures of thousands of degrees and smelts metals and other materials for industry. An even larger solar furnace is being built in France. It is 100 feet high! At present the only American solar furnace the size of the Mont Louis mirror is one operated by the United States Army Quartermaster Corps at Natick, Massachusetts.

A fully equipped solar furnace 5 feet in diameter developing 6,000 degrees Fahrenheit, and suitable for research or industrial use costs thousands of dollars. Larger ones, like that of the Army, cost far more, of course. Few of us could afford even the "small" 5 foot model, but we can build a solar furnace with a lens 14 inches in diameter which is capable of producing a temperature of 2,000 degrees F. or more. That temperature will melt a number of metals and can be used for

Giant solar furnace nearing completion at Odeillo, France, will be world's largest. It is expected to develop about 2,000 thermal kilowatts. Huge parabolic reflector is actually a nine-story building.

DR. FELIX TROMBE

soldering, light brazing, tempering, and annealing. An excellent use for our furnace is the firing of enameled jewelry.

Not nearly so spectacular as solar furnaces, but perhaps of more importance in the long run, are solar homes and sun-heated buildings that are being developed. Heat is a logical thing to get from the sun, but engineers plan to produce the opposite as well; refrigeration systems powered by the sun have shown promise in tests, and solar plants have produced ice in sizable quantities!

The sun is also very effective in the distillation of

This 30,000 square foot solar still, built in the town square on the Greek Island of Symi, produces about 6,000 gallons of fresh water a day.
SOLAR ENERGY SOCIETY

water. Taking a tip from South Americans who operated a solar still long ago to produce thousands of gallons of drinkable water a day in a region having no fresh water, engineers are now desalinating sea water for domestic use in areas of water scarcity. There are a number of solar stills operating on arid Greek islands, for example. Places like Israel and our own southwestern deserts could profit by successful application of such stills.

Our do-it-yourself solar still has a collection area of about 4 square feet. Distillation of sea water—to alleviate shortages caused by the dwindling usable supply and increasing population—will require *acres* of surface, together with special plastic materials with an efficiency of light transmission higher than that of materials now available. In principle, however, the small model we will build is much the same as the pilot plants already built and the full-scale installations to come. It will distill salty or otherwise unfit water, making it suitable for drinking or for use in batteries, steam irons, and so forth, or for experimental purposes.

One commercially available solar-reflector cooker costs about 30 dollars. It is well worth the money and, since it folds compactly, is fine for taking along on picnics and camping trips. Our simpler cardboard cooker is much homelier and doesn't have the advantage of portability. To offset these disadvantages, however, it is much less expensive and quite easy to build. Costing about five dollars, it is made from cardboard and aluminum foil. Yet it will cook hot dogs in ten minutes or boil a quart of water in a little over half an hour, with no time lost to get the fire going!

When you tire of steak, hotcakes, and so forth, there is a solar oven to take care of another kind of cookery. This project is a real oven and uses the "greenhouse" principle to trap the sun's heat. Our oven reaches 400 degrees F. and will bake bread, casseroles, meat dishes, and even pies. It, too, costs only a few dollars.

Solar hot-water heaters have been with us for some time, but today they are better designed. Some swimming-pool owners heat their pools this way for winter swimming. The water heater we will build provides a

Solar water heater mounted on rooftop in Israel.
SOLAR ENERGY SOCIETY

modest supply of hot water for camp or cabin. It uses principles familiar to engineers and architects—those of conduction and convection, and a handy phenomenon called *thermosyphoning.*

So great are the long-range possibilities of solar energy that scientists have called the sun our "peaceful H-bomb." Fortunately, to build solar devices we don't have to be skilled engineers and millionaires as well, as we would have to be to construct a nuclear machine. As a matter of fact, each of the projects presented here is within the mechanical ability of nearly anyone and won't strain even the most modest pocketbook.

The so-called "solar belt" extends from the equator to the fortieth parallels of latitude, and solar devices will work in most of the populated areas of the world. Of course, the more sunshine, the more efficient the device; and the sun high overhead is stronger than when it is low on the horizon. But even as far north as Montreal a solar home has proved its worth. Chances are that unless you are a resident of the polar regions, there will be enough sunlight available to operate the project of your choice.

In addition to the sun projects presented here, there are others that are interesting and thought-provoking. The sundial, for instance, is the earliest form of solar clock. The radiometer, is a sun motor in its simplest form and yet suggests one of the most fantastic space-travel concepts that has yet been considered. There are also electric clocks, cigarette lighters, and hearing aids using solar fuel. All these things, and more, are no longer dreams of science-fiction writers but a reality. In a few days you can have the sun working for *you*, with one or more of the projects in this book.

You will find all of our projects easy. They require no special skills or expensive tools, and their cost is small. Yet, though they are simple, they demonstrate the theories of solar science. In building those of your choice you may well acquire the curiosity to delve deeper into

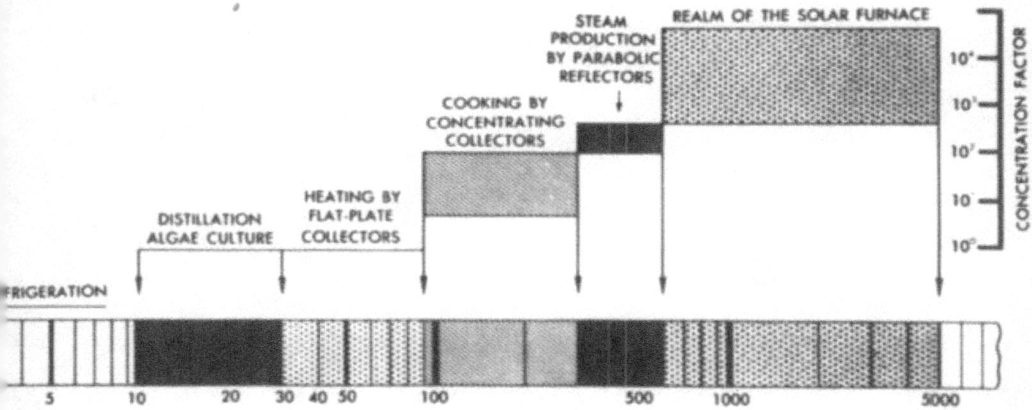

Thermal applications of solar energy.
SOLAR ENERGY SOCIETY

the fields of thermodynamics, optics, electronics, and mechanics. This is the way it should be. Good luck, and have fun with the sun!

2

Solar Furnace

The importance of solar heat is obvious and the expression "a place in the sun" has roots in our vital need for warmth. At times the sun seems to produce too much heat and we swelter and turn brown in its rays. But heat is useful to man in many ways. Since it is the easiest of all the kinds of radiation from the sun to harness we shall begin our projects with a solar furnace: an optical device that increases temperature as lenses enlarge images.

Most of us learned at an early age that a magnifying glass would do more than simply enlarge the image of type in a book. A lens can magnify heat as well; that is, it concentrates the sun's rays into a small space. By merging a relatively large amount of warm rays into a small hot point, we can use the hot point to burn our initials in wood, start a campfire, or, as with our reflector stove, cook a meal. It is this same principle of concentration that makes possible the huge solar furnaces in operation today. A lens or mirror, whichever the case may be, intercepts a certain amount of sunlight and focuses all of its energy onto a tiny spot. This can produce temperatures of several thousand degrees Fahrenheit.

Centuries ago, scientists melted metals at temperatures in the neighborhood of 2,000 degrees F., even with the comparatively crude furnaces they were able to construct. Modern technology has made it possible for the solar furnace to create the hottest temperatures available to man for any appreciable length of time. Other methods such as the "shock tube" and plasma-jet techniques, too complex to discuss here, do produce higher temperatures, but only at great cost and for limited periods of time.

Lavoisier's solar furnace used two lenses instead of mirrors, and reached temperatures approaching 2,000 degrees F.
SOLAR ENERGY SOCIETY

Large solar furnace built by U. S. Army. Used for materials research, it is capable of 5,000 degree F. temperatures and can burn a 4 inch hole in an iron I-beam.
U. S. ARMY PHOTOGRAPH

France's 35-foot solar furnace has operated for many years and has satisfactorily melted metals experimentally and for industrial use. The furnace of the U.S. Army Quartermaster Corps is about the same size and is used for testing materials.

Obviously these furnaces cost fortunes to erect, but anyone can build a workable solar furnace for a few dollars and in a few hours. Far from being a toy, this furnace will melt metals with fairly low melting points, do soldering jobs, and even heat a kiln for firing ceramic jewelry.

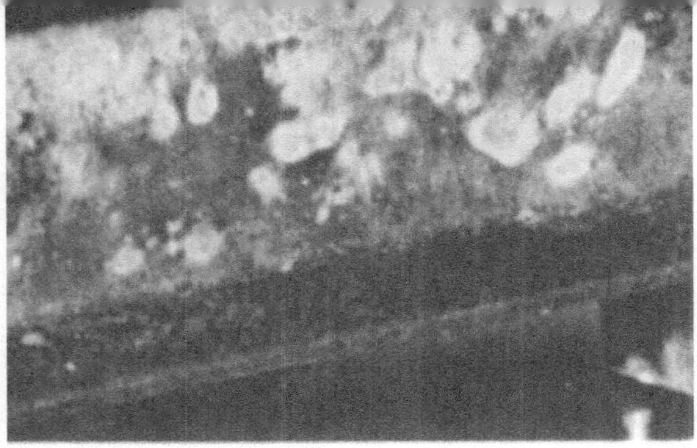

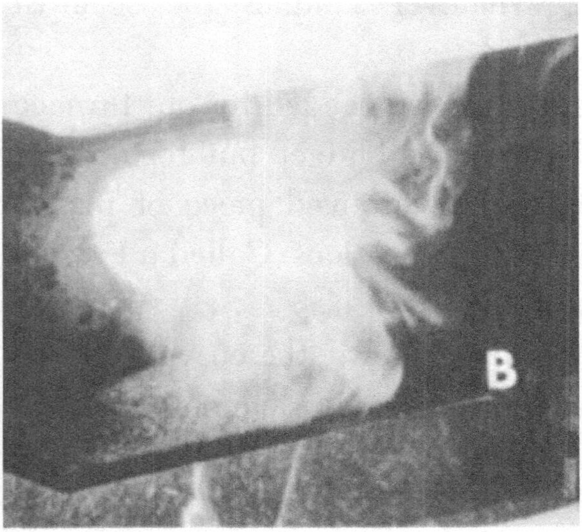

Graphic presentation of the intense heat the Army Quartermaster Research & Engineering Command's new solar furnace can generate is made by these sequence photos taken during an operational test at the Natick (Mass.) site. Sunshine can be concentrated to reach sustained temperatures of nearly 5,000 degrees Fahrenheit.

U. S. ARMY PHOTOGRAPH

In 1770, the French scientist A. L. Lavoisier had convex lenses made especially for him by the great St. Gobain glass works. These curved pieces of glass were joined together at their rims and filled with wine to make them equivalent to solid lenses. Light passing through them was refracted or bent toward a common focal point to heat whatever samples the scientist placed there.

A convex glass lens of the size needed for our furnace would far exceed the average budget but fortunately there is a substitute, a thin, grooved piece of plastic which takes the place of the thick lens. Called a Fresnel lens, this will serve our purpose well.

As the sketch on this page shows, the curved surface of the lens is, in effect, cut into small segments of equal

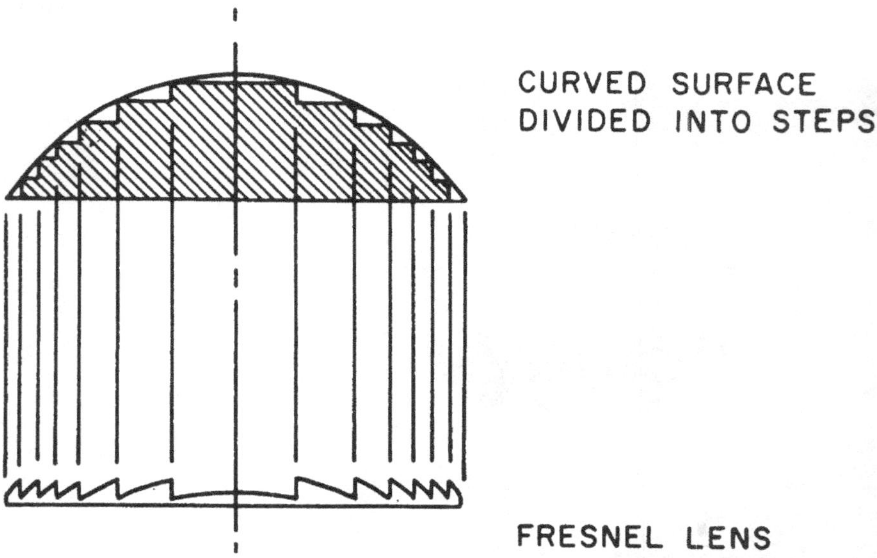

height and flattened out into the thin form shown. Plastic Fresnel lenses are made for a reasonable price from a master form. The one used in our project costs six or seven dollars. It produces about 300 degrees of temperature for each dollar spent!

We could immediately use our lens for melting, soldering, and so forth, much as we would use a smaller burning lens. But holding the lens in one's hands and trying to keep it accurately focused on the work would soon tire the most ardent experimenter. A necessary part of our furnace, then, is the lens mount.

To understand the use of the equatorial mount, consider how the sun moves across the sky during the day. The earth rotates on its north–south axis, and the apparent movement of the sun is at right angles to this axis, that is, it rises in the east and sets in the west. (This is a simplification, and readers familiar with astronomy know the actual movements are somewhat more complicated.) The sun moves along a "parallel of latitude" in its daily journey. We are acquainted with the gradual movement of the sun south in wintertime, but now we are considering just one day's solar motion.

For greatest usefulness our lens should "follow" the sun, and for this purpose the equatorial mount is ideally suited. Such mounts are used in telescopes, so designed that a motor turning at the right speed will keep a star or other body continuously in the field of vision. Large solar furnaces operate in this way or with a photoelectric device that keeps the sun's image properly positioned. Of course, in our simple solar furnace we will not include an automatic tracking system.

The equatorial mount has two axes. One of these is adjusted so that it is parallel with earth's north–south axis. When this is done, movement of the lens on the other axis (which is at right angles to the north–south axis) follows the sun.

Set up correctly, our furnace mount is so designed that we have only to tip the swinging part to follow the sun. In practice we will find that since the sun moves only about one degree in four minutes we can sometimes accomplish our soldering, melting, or other operation without moving the lens. Most often, however, the equatorial mount is more convenient.

Now that we have briefly discussed the theory of operation of our solar furnace let's get busy and build it. See that everything is on hand before you begin. The job will go faster and you will soon have your furnace in operation.

MATERIALS:

1 by 2 inch wood—approximately 20 feet
1 by 4 inch wood—approximately 9 feet
3/4 inch plywood—19 by 19 inches
1 inch wood screws—16
1/4 by 4 inch studs and wing nuts—six
1/4 by 2 inch bolts and wing nuts—four
14 inch (diameter) Fresnel lens—one

1/4 inch chuck—one
Firebrick—one
Window glass—1/8 inch thick; to fit lens (see text)

The furnaces shown in this chapter use a 14-inch diameter lens of about 1/16 inch thickness. Edmund Scientific Corporation of Barrington, New Jersey, sells a variety of sizes and thicknesses of Fresnel lenses in square and rectangular shapes. Select one at least 11 inches square and having a focal length of 16 inches or more. The thicker the plastic, the more rigid the lens. If the plastic is thick enough it may not be necessary to support it with glass as is done in the furnace we are describing. Prices for lenses suitable for our furnace will range from $6 for the 11 inch lenses to $7.50 for one about 12 by 16 inches. Naturally, the larger the lens, the more heat the furnace will develop.

The firebrick kiln may be of any type that will withstand the great heat developed at the focal point.

Use a good grade of wood for the various parts of the furnace. Cheap lumber is a poor economy, because warping will result in a second-rate finished product. In addition to the materials listed, we will need a drill, crosscut saw, keyhole saw, and screwdriver.

On the 19 inch square plywood lens-holder, mark the cutout for the lens. If you plan to use a rigid plastic lens, make this opening slightly smaller than the lens to allow for overlap. If you are using a thin lens sandwiched between glass panes, you can make the cutout

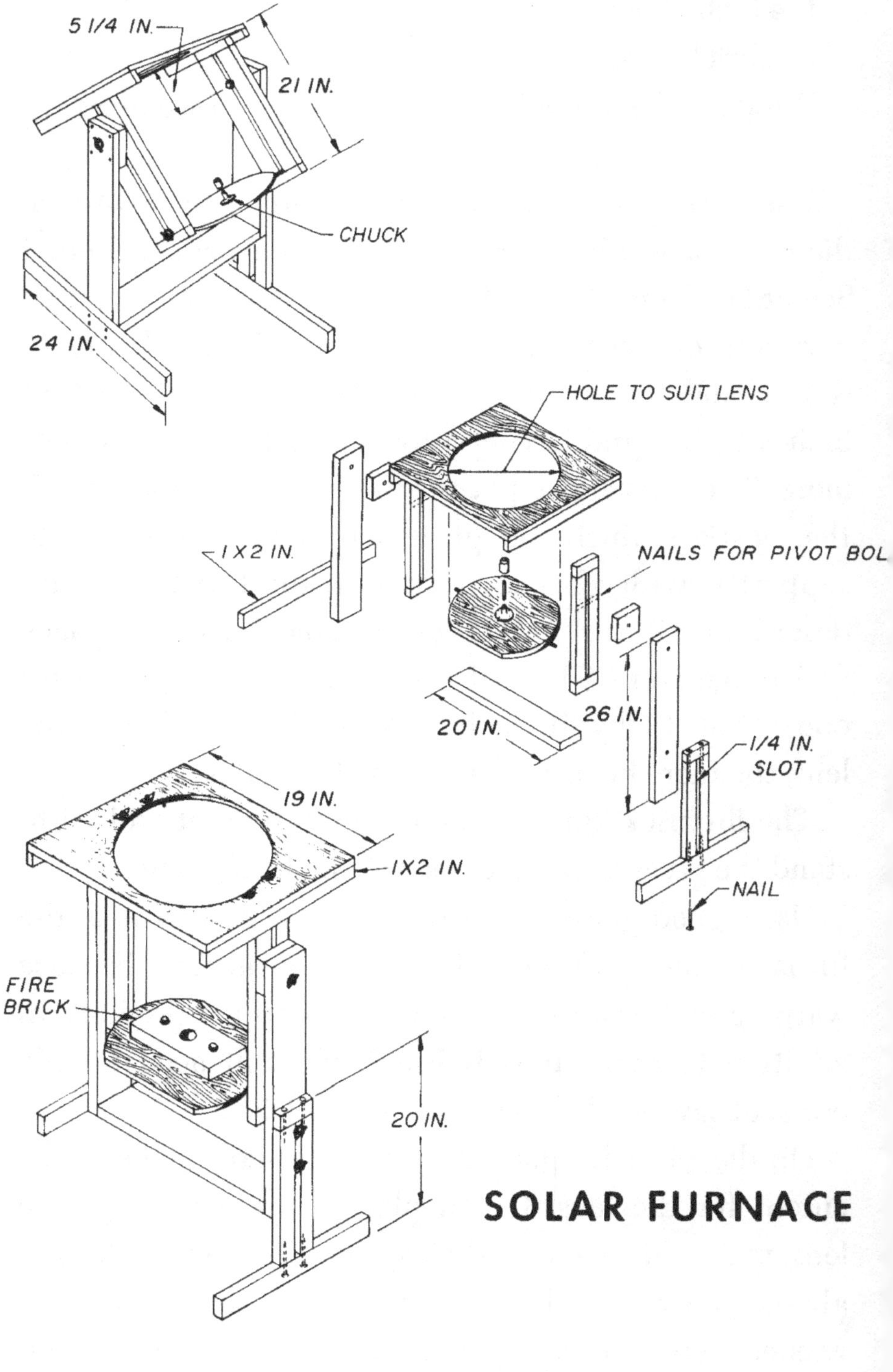

SOLAR FURNACE

the exact size of the lens and let the glass overlap.

Drill a starting hole just inside the line, and use a keyhole saw to cut out the center. Save this piece of scrap for the worktable of the furnace on which we will later mount the kiln and adjustable chuck. Attach 1 by 2 strips to lens-holder, as shown on page 21.

Next cut the 1 by 2 strips that form the swinging frame, making sure all ends are square. Assemble with wood screws to form the slotted track that permits adjustment of the worktable. Drill holes in two places in each assembly for the finishing nails that form bearings for the pivot bolts. Double check to see that the slots are wide enough for the 1/4 inch studs to slide smoothly up and down, and set these parts aside.

Now drill four holes in the plywood lens-frame for the mounting studs as shown in the drawing. Carefully place the slotted parts in position and mark matching holes in their top ends. Drill for studs, and screw the studs in place. Leave enough threads protruding above the plywood for a washer and wing nut, plus several extra threads for mounting the lens in case you use the "glass sandwich" method to be described.

Next cut the three 1 by 4 pieces that form the yoke for holding the swing frame. Note that the dimension on the drawing for the base of the yoke is given as 20 inches. This is suitable for a 14-inch round Fresnel lens. If you use another size lens, adjust the length of the base piece accordingly. Assemble the yoke and nail the square spacer blocks in place on the top, inside surface

Metal sample held in chuck is oxidized by high temperature at focal point of furnace.

of the side pieces. Drill 1/4 inch holes for the pivot bolts through the sidepieces and spacer blocks. Attach the horizontal support with four wood screws so that the yoke will stand upright, and drill two 1/4 inch holes in the sidepiece opposite the horizontal support. These holes are for the bolts that hold the sliding adjustment leg, which is assembled of 1 by 2 strips in the same way as the swinging frame.

Now we can use the plywood piece left from the center of the lens holder. Trim the sides straight to eliminate the starting hole, and then drill two holes on opposite sides as shown in drawing to receive studs. Insert the studs. The worktable can now be attached to the side supports by means of washers and wing nuts.

Place the lens frame in position and slide the pivot bolts between the finishing nails to complete assembly of the mount. All that remains to be done is to install the lens. There are several ways of doing this. It is possible to glue a thin lens in place with Goodyear Pliobond or a similar adhesive. An alternate method is to place the lens between two thicknesses of window glass and attach it to the lens frame by means of the washers and wing nuts.

If you want to use the gluing method, follow these instructions carefully: Remove the plywood frame from the mount and apply glue to the area to be covered by the lens. Several coats may be required to fill the pores of the wood. Add one more coat and let dry for a minute or so. This final coat should be brushed out

Solar furnace shows use of the firebrick

carefully so that it is uniform and not too thick; otherwise the plastic lens will not adhere properly and pull tight, as it should for best results.

Very gently place the lens in position on the frame, keeping it smooth, but without trying to stretch it. Be sure that the grooved side of the lens is up, or the furnace will not operate! With the lens in place, set it aside to dry for a day.

Because the Fresnel lens is plastic and can easily be damaged, a good way to add to its life is to sandwich it between two sheets of glass of the proper size. The photographs show this method. When not in use the lens can be removed and stored for safekeeping. Another advantage is that the lens is held more nearly in a plane than it is in the gluing method. The glass does cut down slightly the amount of heat transmitted through the lens, so the choice is between high temperature and durability.

Some of the Fresnel lenses available are thick enough (almost 1/8 inch) to be quite rigid. If you have this type you may mount it without glue or glass panes, simply attaching the lens with wing nuts as shown.

If you plan to use the furnace only for soldering, jewelry work, and so on, you may want just the firebrick kiln. With a masonry drill, drill two holes in the brick and matching holes in the worktable. If you don't have such a drill, the brick can be wired in place. For jewelry work it is handy to drill or chisel a shallow hole in the brick to hold small pieces.

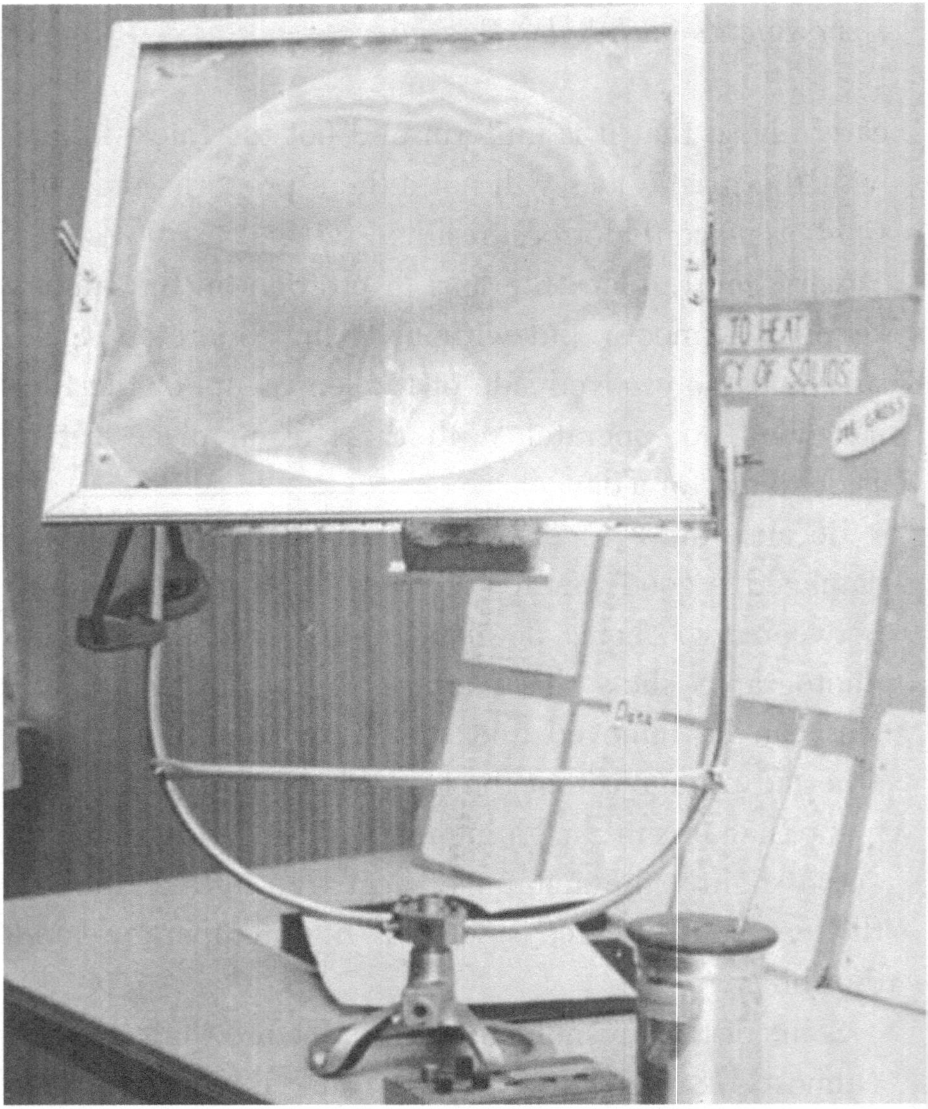

Fresnel lens mounted in simple U-frame. The unit can be used as a furnace.
SOLAR ENERGY SOCIETY

For experimental work it is also handy to have the adjustable chuck setup shown in the photograph. Metal samples are then easily placed at the focal point. A chuck from a discarded hand drill, along with a length of rod threaded to fit the chuck will serve the purpose. A mounting flange is screwed to the worktable to re-

Artist fires ceramic jewelry in solar furnace.
SOLAR ENERGY SOCIETY

ceive the rod, which may be adjusted to position the chuck as required. Both brick kiln and chuck may be left mounted on opposite sides of the worktable.

Using the furnace is simple. Place the mount with the adjustable leg toward the north (unless you live in the Southern Hemisphere), loosen the pivot wing nuts, and tip the lens frame toward the sun. *Wear dark glasses whenever you are using the furnace because the bright spot at the focal point can harm unprotected eyes.* Now move the adjustment leg until the image from the lens is centered on the worktable. It is a good idea to have the firebrick in place when setting up the first time, so that you won't burn holes in the worktable!

Tighten the wing nuts to hold the mount in place,

Samples of jewelry fired in solar furnace.
SOLAR ENERGY SOCIETY

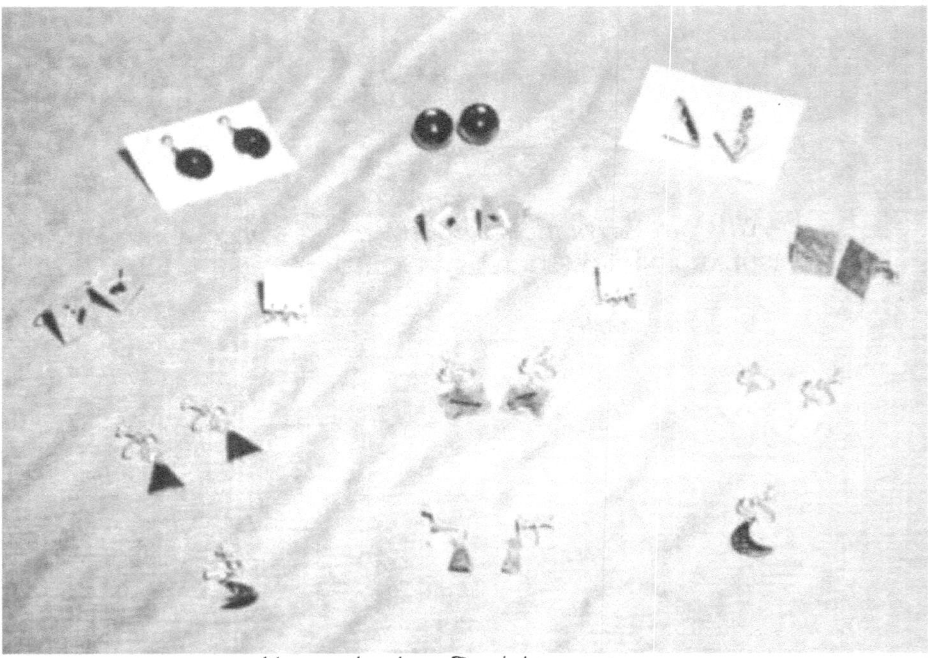

and loosen those holding the worktable. Slide it back and forth until the bright spot in the firebrick kiln is at its smallest. The furnace is now focused, and the temperature should be close to 2,000 degrees F. with bright sun. You will understand now why we have made the table adjustable for purposely shifting focus for less heat. Experience will be the best teacher here. *Meantime, be careful of hands and eyes while you use the furnace.*

If you live in the north and use the furnace in winter, when the sun is low on the horizon, the equatorial mount must tilt considerably. It may be necessary to weight the base of the frame to prevent the furnace from tipping over. For short jobs, you may forget about orienting the base north and south and simply point the lens at the sun.

An alternate solar-furnace mounting makes use of this aiming system rather than the equatorial arrangement. For it we will need only a base plate and a pivoting U-frame. The lens holder is attached to the U-frame with studs and wing nuts. By properly tightening the fasteners, you can make the mount easy to position yet rigid enough to remain in place while you use the furnace. To keep the sun focused you will rotate the U-frame and tilt the lens-holder however much is required.

This simpler mounting will permit soldering, brazing, and similar operations but may require two sets of hands to hold everything in proper position!

3

Cardboard Cooker

OUR FIRST PROJECT was a furnace using a lens. Now we are going to build a stove. Instead of a lens, we use a reflecting mirror that also concentrates the sun's rays sufficiently to create cooking temperatures at its focus. A stove made of paper sounds about as practical as a pitcher carved from ice, but don't be fooled. Constructed almost entirely of cardboard, this reflector cooker will broil steaks, grill hot dogs, fry bacon and eggs, and make hotcakes and coffee. It will also heat water for doing the dishes. No fuel is necessary, because this stove cooks with sunshine!

Stop to think for a minute and you'll realize that every time we cook—be it with gas, electricity, or charcoal—we indirectly use the sun's energy, which has been stored up and reconverted to heat. Basically, then, our solar stove's fuel is nothing really new. Even the use of direct sun heat for cooking goes back many years. People have prepared sun-dried foods for hun-

A solar rice cooker produced by the Goto Company in Japan.
STANFORD RESEARCH INSTITUTE

dreds of years and crude solar stoves were built a century ago. Besides, who hasn't heard of cooking an egg on the sidewalk on a really hot day?

In recent years, however, many advances have been made in the design of solar cookers. Today there are commercial models that are fine for campers or for patio use. One umbrellalike design folds up for easy carrying and storage. Such a cooker is just the thing for trips. If you are dubious about how well the sun can cook a meal, or if you don't have the cash to buy a ready-made stove, get busy and build the one described on the following pages. At most, it will cost five dollars. If you use discarded cartons and other salvage material, the expense will be less.

MATERIALS:

Cardboard–as required
Poster board–two sheets
Aluminum foil–one roll
Plywood–one piece, 18 by 24 inches
1 inch aluminum tubing–approximately 64 inches
1 inch mounting flange–one
Grill–one
Curtain rod–one
1 x 2 wood–four feet
Clothesline–one foot
Glue–as required
Masking tape–as required
3/16 by 1 inch bolt with wing nut–one set

The reflector framework is cut from corrugated cardboard approximately 3/16 of an inch thick, the kind large cartons are made from. Some poster board and aluminum foil will complete the cooker itself. A grill stand (for hot dogs, hamburgers, or pans) is made from plywood, some tubing, and an inexpensive hand grill that costs about 50 cents.

Study page 37 first to get the over-all picture and to see how much new material will be needed. If you want to buy new cardboard, two 4 by 8 foot sheets will be plenty. These cost about 80 cents each at a box factory or supply house. The other items will be easy to find. Get all the materials ready and then begin construction. An eager beaver can do the job in a day or so and begin sampling outdoor cooking à la sun right away.

First, cut a base piece 3 feet square from the 3/16-inch cardboard. Mark the layout of the reflector ribs right on this base. This is the size our finished cooker will be. Next, draw two diagonal lines through the center of the base, perpendicular to each other, as shown on the plans. These mark the location of the main ribs, which we will make next.

A word about the principle of our reflector cooker will be helpful before we proceed. The sun stove focuses the sun's rays that strike its surface onto the bottom of the grill. Even on a clear winter day our cooker collects a lot of "warmth," which when shrunk into the 1 foot area at the grill, becomes concentrated "heat."

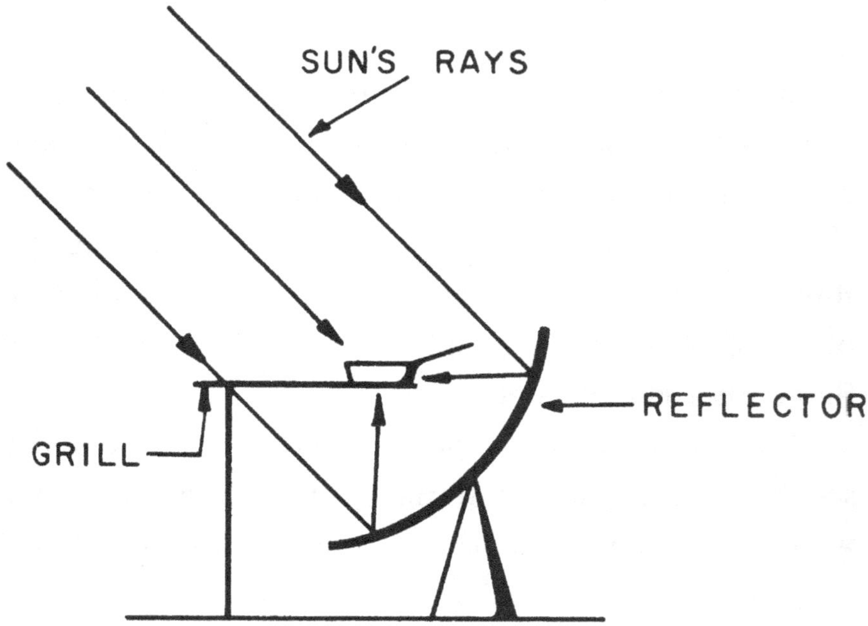

How solar-reflector cooker works.

The giant solar furnaces we talked of in the first chapter use curved reflectors too. They generate thousands of degrees at their focal points, using the same principle. To do this they must be of very accurate parabolic shape. This specially shaped curve reflects all the rays onto one tiny spot and gives the furnace a concentration ratio of many thousands to one. Obviously we don't want such high temperatures, for they would melt our pans!

Our reflector uses a radius of 36 inches instead of a true parabolic curve, resulting in a larger spot at the focal point. We will also use a number of wedge-shaped sections instead of one bowl-shaped reflector. Thus our focal spot will be roughly the size of the cooking pan, which is just what we want.

Now that we understand what we are doing, let's draw two main ribs as shown on the plans. To draw an accurate radius on the cardboard, drill two holes 36 inches apart in a strip of wood something over three feet long. Make one hole for a nail and another large enough to slip a pencil through. Drive the nail into a piece of wood that is firmly anchored and swing an arc with this king-size "pencil compass," making sure the plywood rib material is anchored down, too. Use this method on both full ribs instead of tracing one from the other, since the latter process may well result in discrepancies. Cut these ribs carefully, using a sharp linoleum knife, pocketknife, or model builder's razor knife. Be sure to plan ahead so as not to waste material as you lay out the ribs. Each of the main ribs has a notch at the center. Notice that one is on the top and one on the bottom so the ribs will interlock.

Using a full rib as a pattern, mark out twelve partial ribs, as shown on the plans. Before cutting these, cement the full ribs to the base plate on the lines previously drawn. Model airplane glue or a good household cement will work well. While the parts are drying, cut out the remaining ribs.

Notice that in addition to the full ribs which fit diagonally on the base plate, there are two other sizes. We will need four of the short, partial ribs and eight of the longer ones. These have the same 36 inch radius, but their inner and outer ends are trimmed to fit in the spaces allotted them on the base. Be very careful in

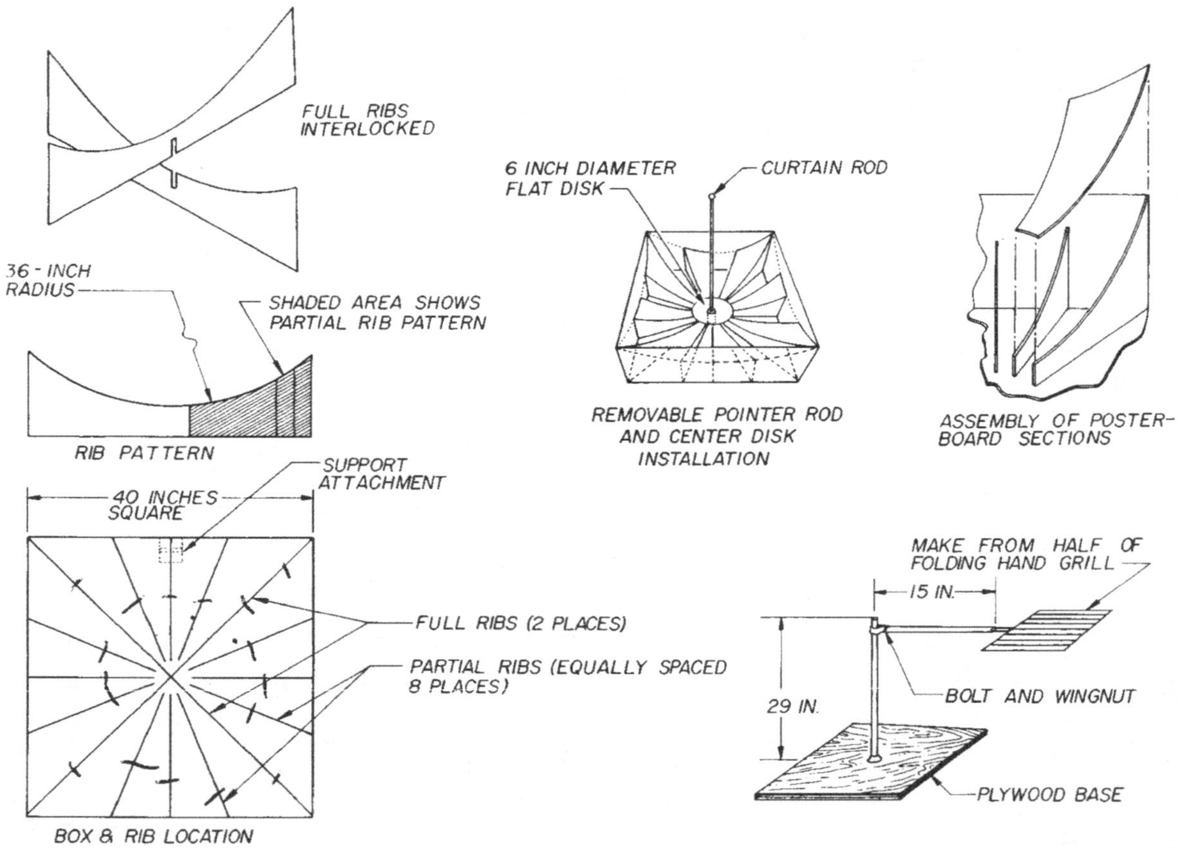

REFLECTOR SOLAR COOKER

cutting out the ribs. The effectiveness of the finished cooker will depend on your accuracy in cutting and installing the ribs, so take your time and do a good job.

When the framework is thoroughly dry, you are ready to put on the wedge-shaped pieces of poster board. Since these form the curve that will reflect the sun's rays, use poster board that is thin enough to bend easily, yet has sufficient body to hold the proper shape. Lighter cardboard would ripple and wave.

By means of cut-and-try methods, trim one piece of poster board so that it covers the space between two ribs, with about a 1/8 inch overlap all around. Do not cement this in place yet; it will be your pattern for more pieces. Cut them carefully, making sure they will cover any of the spaces between ribs. (In spite of care, there may be slight inaccuracies in the framework.) It is better to have the poster-board pieces a bit too large than too small.

There are two sizes of wedges to form the curve of the reflector. Make a pattern of each, and then cut out the others from these. Cut eight of each size. With all the pieces cut, now begin to cement them in place. Since butting the joints smoothly against each other would be difficult, glue eight pieces in alternate spaces first. Spread glue along the tops of two ribs, then lay the poster-board wedge in place and carefully press down so that it touches the ribs at all points. The glue will dry well enough in a minute or two so that you can go on to the next piece. Don't forget to leave every other section open.

Ribs installed in box will support reflector.

Half of the wedges are glued in place in the next stage of construction.

Completely covered reflector.

When these joints have dried, cover the open spaces with the remaining eight pieces of poster board. These will slightly lap over the edges of the pieces already glued in place, thus making a strong joint. At the center, where all the points come together, simply trim them off an inch or two. The hole left will later be covered with a separate piece of poster board.

It is possible to cover the reflector with aluminum foil from the handy kitchen roll of this material. But it is much easier to use foil already joined to a heavier paper backing. That used in the cooker described is aluminized Mylar plastic purchased from Edmund Scientific Corporation. Because of the strong paper backing it is easy to apply the foil smoothly. Use a cement of your choice: contact cement, household glue, rubber cement, or what have you. Again, the better the job you do the more efficient and durable your cooker will be.

Now install a marker for the focal point of the reflector so that you will know where to place the grill for the fastest cooking. This marker is simply a small, inexpensive curtain rod of the type used on kitchen doors. It consists of two tubes, one fitted inside the other. Cut a short length of the larger tube and insert it into a hole punched in the center of the reflector where the ribs come together. Better still, use a drill the same size as, or slightly smaller than, the tube to give a snug fit. Now cement the tube in place.

The smaller tube will fit into this "holder" and can be removed for easier handling when not needed. As mentioned before, the focal point of the reflector is the proper place for mounting our grill. With a spherical reflector the focal length is half the radius; in this case 18 inches. As a double check, aim the reflector at the sun and adjust the tilt until there is no shadow visible from the pointer rod. Then hold a piece of wrapping paper with a small hole punched in it right at the tip of the pointer. Move the paper toward the reflector and then away from it until the smallest spot is observed on the paper. This is the actual focal point, and the pointer rod should be cut to this length.

Cut two small squares and one rectangle of cardboard, as shown by the dotted lines on the plans, and cement them to the back of the cardboard base. The squares go first, and then the rectangle. After these are

Detail of the reflector support; nail in dowel engages loop attached to cooker.

well dried, run a short length of clothesline through the slot and tie the ends in a square knot. Drill holes through the 4-foot length of 1 by 2, spacing the holes about an inch apart halfway down the dowel. Insert a nail to engage the loop of clothesline. We can now set up our reflector so that it will stand alone.

To make the grill, first cut a plywood base 18 by 24 inches. Any thickness from 1/2 to 1 inch will do. Mark the center and install a mounting flange for the 1-inch

The reflector cooker in use. Rod indicates point at which heat is concentrated.

aluminum-tubing vertical support, which is 28 inches long. The adjustable arm is also aluminum tubing, 20 inches long. First, flatten one end and bend around a piece of pipe or a broomstick to make the collar, which fits over the vertical support. Drill a 3/16 inch hole in order to insert a bolt and a wing nut. The other end of the adjustable arm may now be flattened. Be careful to keep the flat area at right angles to the collar so that the grill will be horizontal when installed. Slide the grill in place and the solar cooker is complete.

Now that the work is done the fun starts. Positioning the reflector is simple if you follow these directions. Stand behind it and face it toward the sun. Now tilt it until the shadow of the pointer rod vanishes as it did when you checked for focal length. This means that the reflector is aimed properly and that the sun's rays will be bounced right where you want them.

Holding the reflector in this position, slip the 1 by 2 support through the rope loop and put the nail through the hole just below the loop. With the reflector on its own feet, you can now put the grill in place. Loosen the wing nut on the adjustable arm and move it up or down until the grill rests just above the tip of the pointer rod. As a double check, pass your hand quickly just above the grill. It should be hot, ready for you to start cooking.

The grill surface itself is fine for cooking hot dogs, burgers, or steak. Grease will drip on to the reflector but will not harm it. For bacon and eggs, hotcakes, and the like, place a skillet on the griddle. And if you like

Cooking hot dogs over larger solar device.

your steaks seared quickly to keep in the juice, use the skillet for them, too. By putting it on the grill a few minutes early you can store up extra heat that will cook the steak more rapidly.

Water for coffee, tea, or for dishwashing can be heated in a kettle or pot. To get the maximum efficiency from your solar cooker, use blackened utensils; however, just about any kind will work satisfactorily. For variety try a pressure cooker.

As the sun moves across the sky, the position of the reflector must change. In early morning or late afternoon it will be nearly vertical, while at noon in midsummer you may have to place it nearly flat on the ground. That's why we drilled so many holes in the support rod. If you plan to boil beans or make stew, occasional adjustment of the reflector will be required to keep the hot spot where it will do the most good. The shadow from the pointer rod is the thing to watch. For bacon and eggs, hot dogs, and even steak, one setting will often do the trick.

After cooking your meal and washing the dishes, remove the grill from the aluminum tube and clean it, too. Then wipe off the reflector surface with a paper towel or damp cloth and that's all there is to the job of solar cooking.

Solar stoves won't take the place of other kinds of cooking all the time. When the sun goes down you had better be through cooking, and on a rainy day the reflector is not much use except maybe to crawl under to keep dry! But properly used in clear weather it will amaze the most skeptical observer. Here are a few of the advantages of solar cooking.

As you discovered when you held your hand close to the focal point, there is no warming-up period with a solar stove—it is hot right away. By the time the fellow with the charcoal brazier gets a good bed of coals, you will be doing the dishes. Besides, he paid for his fuel while yours was free for the taking. And solar energy is available any place the sun shines: mountains, desert, beach, or your own backyard.

At first, the reflection from your cooker might be bothersome, and a pair of sunglasses will be handy. After practicing a while you'll learn where to stand so there isn't any glare, and by then you will have noticed how nice it is not to have your eyes full of smoke. Solar cooking is cool cooking, too, because the heat goes into the food on the grill and doesn't roast the person doing the cooking as well.

You won't need matches to get your cooker going, and there's no danger of setting anything on fire either. Since the solar cooker produces no open flame it may be used safely even in areas where cooking fires are not permitted because of dry conditions or other hazards. Lastly, there are no ashes or soot to contend with.

You will have a lot of fun cooking with sunshine. It's safe, it's clean, and it's free. Chances are you'll find it portable enough for your next camping trip. That way you won't be tied down to a fireplace and the bother that goes with it. So save up for a commercial folding cooker, or you might even put your ingenuity to work and build a version of the cardboard stove that folds to make it more portable.

4

A Solar Oven

THE "GREENHOUSE" effect is well known to those who grow plants in such structures and also to those who have left the windows of a car rolled up on a warm, sunshiny day. The result is aptly described as resembling an oven. That is what we are going to build in this chapter—a solar oven that will do a real job of cooking on a clear day, even in winter.

Cooking with sunshine is an ancient idea. Man has dried fruits and meat in the sun for ages, a kind of leisurely cooking with solar energy. Direct use of sunshine for cooking dates back more than a century. When Napoleon was looking for a convenient way to cook for desert troops, a French scientist named Augustin Mouchot produced a solar cooker that did the trick nicely. In our own country, Dr. S. P. Langley experimented with solar cookers atop Mount Whitney many years ago. Both these men used the so-called "hothouse" principle. A glass cover on a box admits heat

Solar scientist Dr. Maria Telkes demonstrates use of an aluminum cooker of her design.
SOLAR ENERGY SOCIETY

readily, then doesn't let it get back out. This is because the incoming heat is "shortwave" radiation and passes easily through the "optical window" of the glass, while the "long wave" reflected heat waves are blocked by the glass. The cover on a solar oven is thus a kind of one-way valve for solar energy.

In recent years the solar oven idea has been revived and suggested as a means of cooking in undeveloped areas short of fuel as well as for campers and hobbyists. However, most solar ovens are of metal and quite heavy and expensive. The author's original oven was of

How solar oven traps heat.

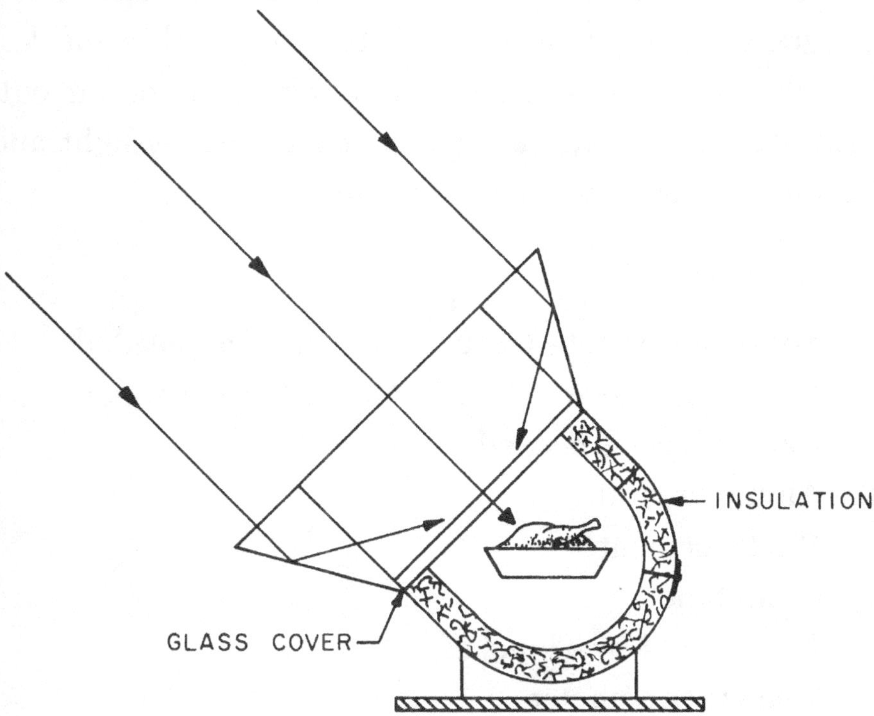

this type, built of galvanized iron insulated with fiberglass. It cooked nicely, but was considerably more complicated to construct. The oven presented here takes care of such objections. It is light, compact, cheap, and very easy to build.

Our second project was a solar cooker made of cardboard. A solar oven is even more practical, particularly since it is more compact and not only cooks meat and vegetables, but bakes bread and cakes as well. We could make an oven from cardboard too, but for variety let's use another unlikely material. How about styrofoam plastic? Put a chunk of this stuff in the kitchen oven at the temperatures we hope to reach with our solar oven and in a few minutes all you'll have left is a little ball of melted plastic. So how can we bake at 400 degrees F. in a plastic oven? It's easy with some foil insulation inside and the cooling effect of the air outside the oven. The result is a cooker that is light and compact enough to take just about anywhere.

MATERIALS:

Styrofoam sheet–1-1/2 inches thick as needed
Glass–one piece, 1/8 inch thick, 13 inches square
Cardboard–as needed
Aluminum foil
Plastic cement
Cloth tape
Dowel–1/4 by 48 inches
Oven thermometer

The body of the oven will be made of foam plastic about 1-1/2 inches thick. Buy a sheet of this material from a local supplier and you'll have enough for a couple of stoves. With a wood saw cut out two identical side pieces, a top, bottom, and back. Cement the pieces together with an adhesive that won't dissolve the plastic. The plastics dealer can provide this, or you can use Wilhold or a similar product.

When the oven body is assembled, add four strips of 1/8 by 1 inch balsa wood, as shown on the plan. These strips form the seat for the glass cover which is also the oven door. This is ordinary 1/8 inch window glass, hinged with cloth tape to the oven, and fitted with a tape handle for opening. Have glass cut to size and the edges beveled for safety.

Now we have a primitive form of the solar oven. As mentioned earlier, the pioneer designs by Mouchot and Langley were just boxes fitted with glass lids. The heat of the oven depends on the amount of radiation that gets into it, of course. By increasing the size of the glass-covered box you can collect heat. But there is another, simpler, way of doing the same thing. With reflector panels, which double as protective covers for the glass, you can bounce more solar radiation into the oven.

If we mounted a reflector at a 90-degree angle to the glass it would not do any good, since solar rays would go right by it. Suppose we mount the reflector at a 45-degree angle? Now the rays bounce off all right, but

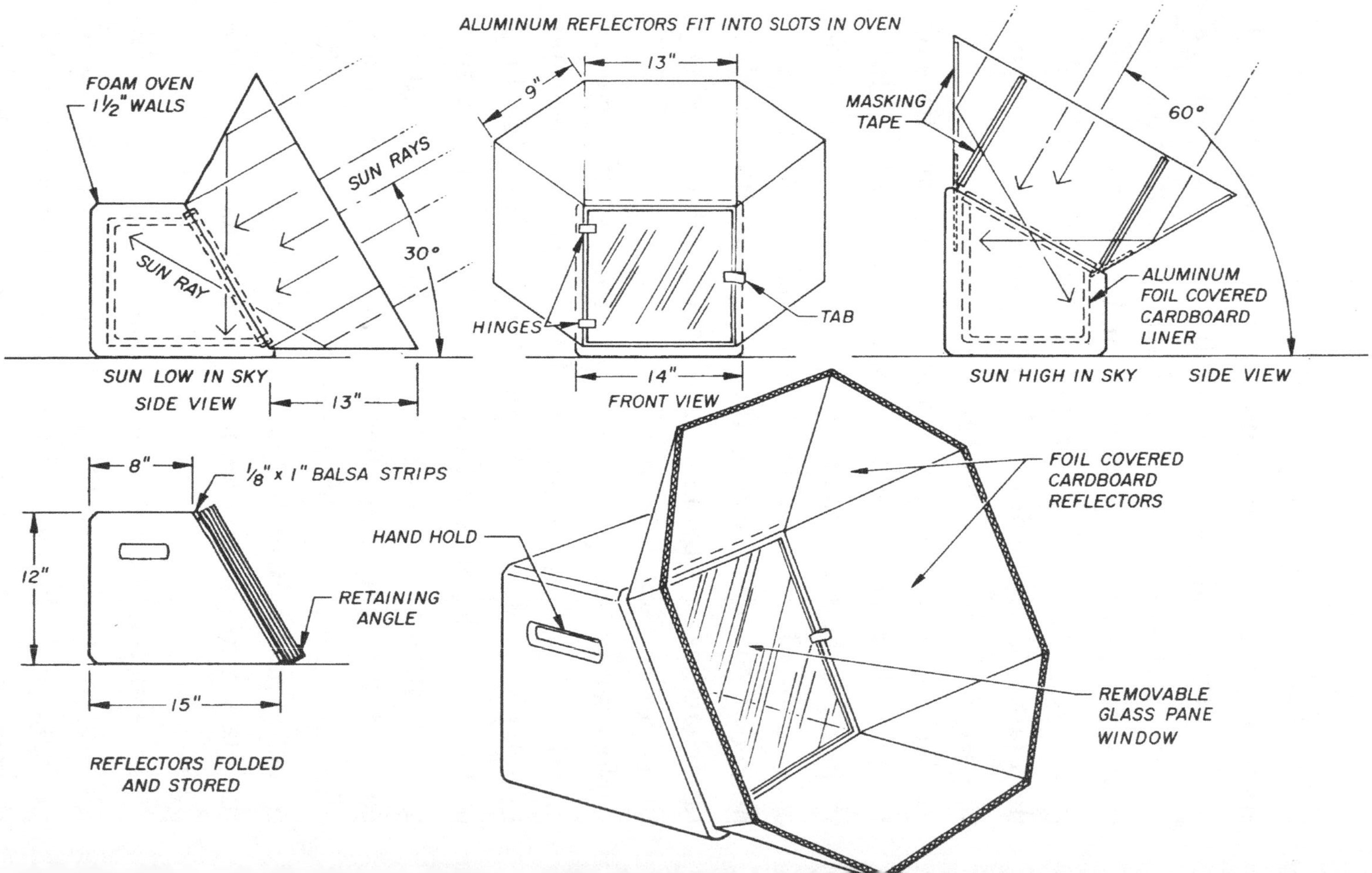
ALUMINUM REFLECTORS FIT INTO SLOTS IN OVEN
FOAM OVEN 1½" WALLS
SUN RAYS
30°
SUN RAY
SUN LOW IN SKY
SIDE VIEW
13"
9"
13"
HINGES
TAB
14"
FRONT VIEW
MASKING TAPE
60°
ALUMINUM FOIL COVERED CARDBOARD LINER
SUN HIGH IN SKY
SIDE VIEW
8"
1/8" x 1" BALSA STRIPS
12"
RETAINING ANGLE
15"
REFLECTORS FOLDED AND STORED
HAND HOLD
FOIL COVERED CARDBOARD REFLECTORS
REMOVABLE GLASS PANE WINDOW

they don't enter the oven. Instead they move parallel to the glass cover. This doesn't help either, but we are closer to a solution and all we need is the correct angle. If you're good at geometry you'll quickly arrive at the correct angle, 30 degrees from the vertical, or 120 degrees from the surface of the glass. At this angle, a reflector the same dimensions as the oven glass will bounce the radiation hitting it into the oven—just what we want.

A little more geometry will show us how much heat we have added with our one reflector. With the reflector at an angle of 30 degrees we add 50 percent to the amount of heat the oven alone gathers. If we add two reflectors, we double the heat. With four reflectors, and sections to fill the corners too, we can add two and one-half times the original heat of the oven.

Cut from cardboard four main reflectors and four triangular pieces to the dimensions on page 53. Cement on aluminum foil. Cloth tape serves as hinges for the triangular pieces and the main reflectors are mounted to the oven with dowels cemented to the reflectors, as shown. Simply push the dowels into the cardboard corrugations, liberally coated with glue, and let dry. Next drill or press holes into the oven itself at the proper angles to receive the dowels.

Line the inside of the oven with cardboard coated with "rippled" aluminum foil. This foil serves as insulation, while the wavy design cuts glare. Don't be tempted to use the "ripple" foil on the reflectors, how-

Compared with a conventional solar oven made of metal, this foam model is very easy to build and actually does as good a job or better. To use it, all you need do is set it on the ground in the sun and load it with food in the proper baking dishes or pans. Note that the shape of the oven is such that you have two choices of reflector angle. When the sun is low in the sky place the oven so that its glass cover is close to a right angle to the sun. When the sun is higher, set the oven on its "back" to catch more sun and heat. The glass door works easily in either position. When not in use, the cooker should be dismantled and the reflectors stored in the rack. This also protects the glass cover.

Left: The finished cooker is compact and light enough to be easily carried. Right: The glass cover of the oven opens easily to permit loading of food.

In case you can't find foam in bulk, you might try using a foam icebox, cut off at the proper angle and with a glass cover added. There are also other ways of making the reflectors. One of these is to use polished aluminum instead of cardboard covered with foil. Hinge the triangular corner pieces to the main reflectors as we did earlier. Then hinge the main reflectors to the body of the oven. Use cloth tape. Now the reflectors can be folded flat against the glass cover, protecting it and making the oven very compact. Fold bottom reflector first, then sides, and finally lower the top reflector.

To guard against the cooker being tipped over on a very windy day, add some wire U-shapes (like croquet wickets) that slide out from the bottom of the oven and can be weighted to anchor the oven.

The ovens described reach 350 degrees F. on a clear winter day and over 400 degrees F. in summer. To cook more food you can scale up the box and reflectors to half again the size shown. With a little effort and ingenuity a rack may be added to make more cooking space. The solar oven will bake bread, cook meat and vegetables, casseroles, and desserts. In a pinch you can use it to heat water too. It is even safer than the cardboard cooker; just don't burn your fingers on the glass door!

5

Water Heater

So far we have used solar heat to melt metals and to cook food. We can also heat water with solar energy. Such heaters have long been used, and as early as the 1930's southern California and Florida abounded in glass-covered heat collectors containing heating coils and a storage tank for hot water. Perhaps you have seen a solar water heater mounted atop a home.

The crude, makeshift designs of earlier days did a fair job of providing free hot water for domestic use, although standby heat sources were necessary for cloudy days or nighttime use. Today there are thousands of scientifically designed and highly efficient solar water heaters—particularly in Japan and Israel, where they compete successfully with gas and electric heaters.

In our solar oven we trapped heat inside the glass-covered box and then transferred that heat to the food we wanted to cook. In a water heater the heat transfer

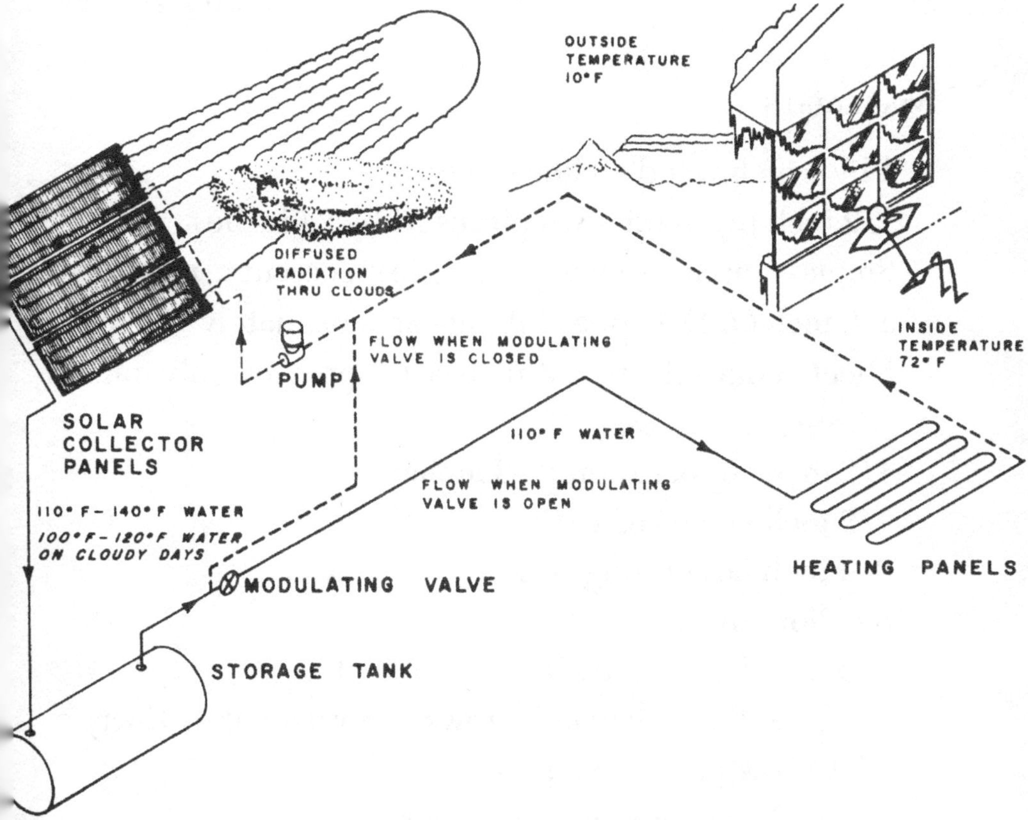

Large solar water heater can create enough heat to keep a building warm.
BRIDGERS & PAXTON

is to water. The sun's rays warm water even without a special heating device. By adding a coil inside our hot box and circulating the water as it is warmed, we can make a more efficient water heater.

The heater described in this chapter is not intended for domestic use, but the 5 gallon tank would make a good supply for camping trips or for a cabin that has no provision for hot water. If desired, an enlarged version of the heater could be installed on a roof, connected to a water supply, and used as a permanent hot-water source.

MATERIALS:

1 by 4 inch wood—8 linear feet
1/2 inch plywood—two pieces 24 by 24 inches
Single-weight window glass—one piece, cut to measure
3/8 inch O. D. copper tubing—approximately 16 feet
Sheet metal—22 by 22 inches (copper or galvanized iron)
1/2 inch copper tubing—3 inches
1/2 inch valve—one
3/4 inch hose fitting—one
5 gallon can—one
3/8 inch I. D. plastic tubing—10 feet
1/4 by 1-1/2 inch wood screws—approximately thirty
Flat black paint—one pint
1/4 inch stud and nut—one each
Carrying handle—one

Begin the heater with the collector box itself. Make the sides of the box from 1 by 4 material. At the lumberyard have a groove cut 1/8 inch wide to a depth of 3/8 of an inch in the 1 x 4 board. Locate the cut 1/2 inch from one edge. This is the slot for the glass window. The next step is to cut the pieces to the proper length, being sure to keep the ends square both ways. The glass should fit snugly. Cut out a square of plywood 24 inches on a side and place the 1 by 4 pieces on it to make sure they fit. Assemble the sidepieces to the back with wood screws.

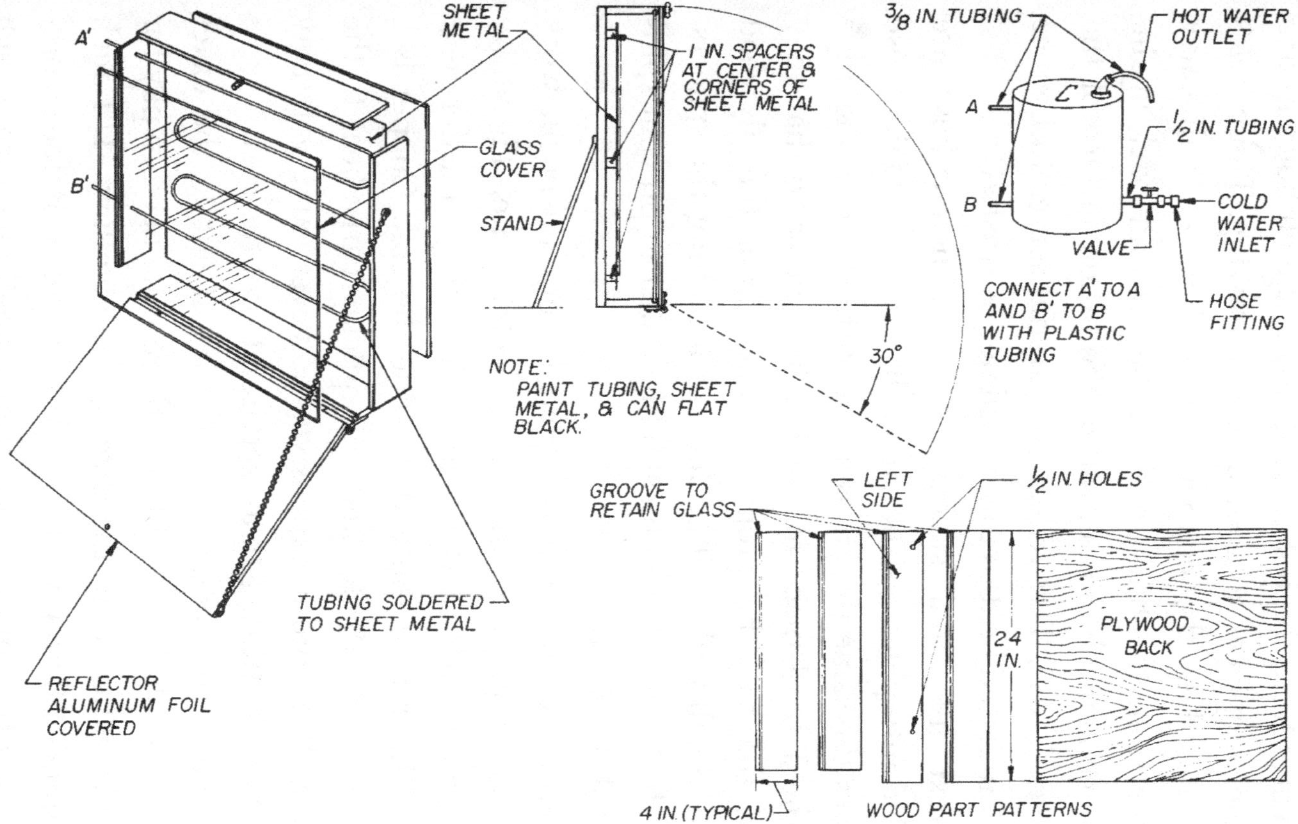

SOLAR WATER HEATER

Take the box apart at this point and clean chips and shavings from the holes. Drill two 1/2 inch holes in one of the sidepieces, as shown on the drawing. These holes accommodate the copper tube coil which will carry water from the tank to the collector and back again. Now cement aluminum foil to the inside surface of the plywood base. This reflective material serves to bounce back radiated heat so that it will not be wasted.

Next, five small spacer blocks (1 inch cubes) are nailed into place, as shown on the drawing. Make these blocks the same thickness as the distance from the edge of the 1 by 4 piece to the 1/2 inch holes so that they hold the coil the proper distance from the plywood base. Small nails will be fine for attaching the spacers, but drill a hole through each block first to prevent splitting of the wood.

With the box itself completed, you can begin work on the copper coil and the collector plate. These are important parts of the heater, since they transfer heat from the sun to the water inside the coil. Copper is a very good conductor and will quickly carry heat to the water. It is quite expensive, however, and you may want to substitute galvanized iron to cut the cost.

If you use a copper sheet, have it cut to the exact size. It is sold by weight, and there is no need to pay for scrap. Notice that the size specified allows 1/4 inch clearance all around the inside of the box. Before working with the copper sheet, trim a small piece from each corner at a 45 degree angle to prevent being cut by the sharp edges.

Bend the heating coil from 3/8 inch copper tube, the flexible kind that comes in a roll. The length called for in the list of materials allows for trimming. First straighten out the tubing, making it as flat and true as you can. It is quite soft, and a little time spent should result in a smooth job. Now lay the tube across the flat sheet of copper, with the proper length extending beyond the edge. Mark the start of the first bend in the tubing with a pencil and then carefully form it with your fingers into a U-shape. Work slowly and evenly so that you will not flatten the tube excessively.

After the first bend is made, replace the tubing on the sheet and make sure the bend is in the proper position and that sufficient tubing extends past the edge of the sheet. Mark the second bend and proceed as before.

Assembly of heat-collector tubing into foil-lined box.

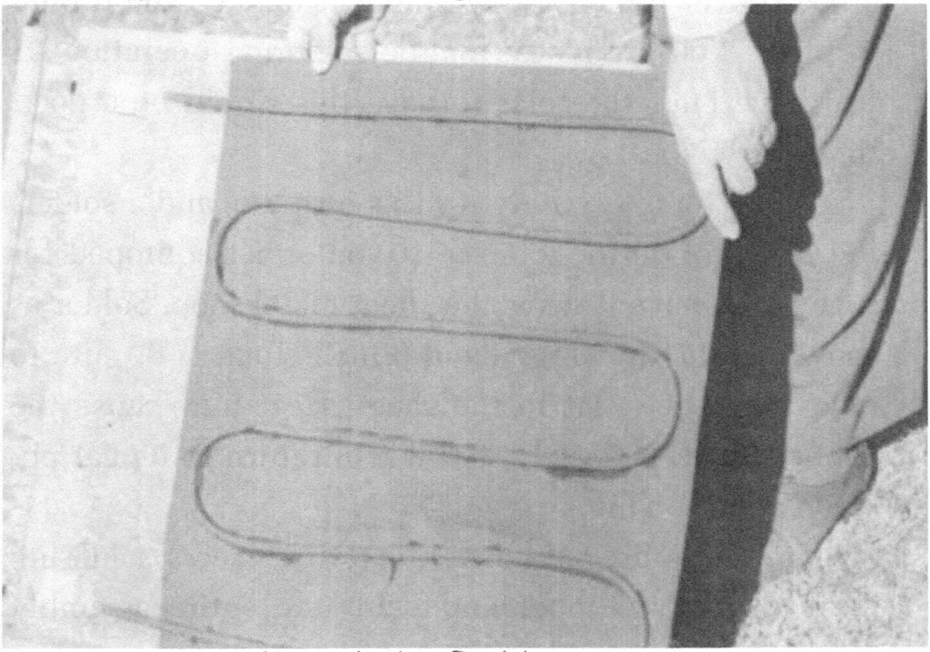

Continue to form the coil in this manner until the sheet is covered in a series of S-turns, as shown.

Trim the long end of the tubing; check the shape of the coil once more; and then lay it on a flat surface to see that it is level. Spend as much time as required to make the tubing lie perfectly flat, using your fingers and tapping lightly with a rubber or wooden mallet for the finishing touches. The tubing should touch the sheet along its full length for good heat transfer.

When you are completely satisfied with the job, solder the copper coil to the sheet. Clean the tube and the sheet with emery cloth so that the solder will stick properly. Lay the sheet on a wooden surface (the inverted collector box itself will do nicely), and place the coil in position. Remember that the ends of the tubing must fit through the holes drilled in the 1 by 4. Now lay a board over the coil, and weight it to keep the tubing in place. You are ready for the soldering operation. If you have built the solar furnace this would be a good time to try it out!

A small torch is handy for this purpose, and a soldering iron will do the job too. If you aren't equipped for such work, have it done at a sheet-metal shop. Solder as shown on the drawing, about 6 inches apart. Be sure to hold the tubing flat to the sheet. Heat may cause the copper to warp slightly, but it will return to its flat position upon cooling.

With the job completed, clean any excess soldering paste from the copper and paint the entire assembly flat black. Apply a second coat of paint for good measure and set it aside until it is completely dry. The coil assembly may then be slipped into the box, with the ends of tubing carefully inserted into the 1/2 inch holes. Tack the copper sheet to the spacer blocks when you are sure it fits properly and will not have to be removed.

Unscrew the top 1 by 4 piece. Slide a piece of cardboard into the grooves. Trim to fit and have your glass supplier cut to this measure. Fit the glass into the slots, and replace the top piece. When you install the carrying handle, the collector is complete and you can begin work on the water tank.

A round, 5 gallon can with narrow, screw-top spout is used for the storage of heated water. The one in the photograph was a discarded oil can obtained from a local distributor. Other types of containers are suitable and may be substituted if the round type shown is not available. For example, a square, lightweight can will fill the bill. This type is usually on sale at hardware and surplus stores.

Clean the can of any residue of oil or other liquid. This is done for two reasons. First, we don't want the water contaminated and, second, heat from soldering operations might set fire to the liquid. So do the cleaning carefully and flush with water several times.

Two short lengths of 3/8 inch copper tubing are soldered to the can. The location and dimensions are shown in the drawing on page 60. First drill a 1/4 inch hole in each place a tube is to be installed. Next drive a center punch or other tapered piece of metal into the hole. This enlarges the hole and also forces the metal inward. Check frequently during this flaring process to insure a snug fit of tubing. The depression formed in this way will hold more solder and make a stronger joint.

If a painted can is used, it will be necessary to scrape the areas where soldering is to be done. When the metal is clean and bright, insert the tubing (which has been cleaned too). Using a solar furnace, torch, or soldering iron, let solder flow into the depression and around the tube. This operation is easier if the can is positioned with the tube pointing straight up.

The hot-water outlet is also a length of 3/8 inch tubing soldered to the screwed-on cap of the can. Use the "flaring" method again so that a strong joint will result. Notice that the tube is bent into a U-shape.

You are now ready to do the plumbing for the cold-water supply line. Instead of 3/8 inch tubing, use 1/2 inch tubing for this connection. Attach a simple shutoff valve to the tube, using compression-type fittings that come with the valve. Your dealer will explain how these fittings are installed. Another short length of 1/2 inch tubing extends from the valve. The free end of this tubing is soldered inside a brass garden-hose fitting of the type used with a plastic hose.

With the soldered joint made and the compression fittings tightened securely, you can connect the tank to the end of the garden hose and check for leaks. Turn on the water at the faucet; open the valve at the tank; and cork the tubes that will lead to the collector. When the tank is full, water will overflow from the hot-water supply outlet. Mark any leaks, drain the tank, and repair as necessary. The tank is now ready to be attached to the collector coil.

Your heater would be of little use if only the water in the coil itself became hot; this would barely be enough to wash one's hands. If you made a very large collector, the coils would hold ample hot water. Another method would be to install a pump to circulate water between coils and tank. This would cost more money and also make the heater more complicated. Fortunately, there is a phenomenon called thermosyphoning, which will heat the whole tank of water. Thermosyphoning is the ability of water to circulate of its own accord when heated—given certain conditions. The most important of these conditions is that the supply tank be located *above* the coil.

The photograph shows the tank mounted on a stand, with the bottom of the tank about on a level with the top of the collector. As the coil heats water inside it, this water rises and is replaced by cooler water drawn from the bottom of the tank.

Using the hinged prop, set up the collector facing the sun. Then place the tank on its stand to one side. Install the plastic hose between the two bottom tubes, using hose clamps for a watertight connection if necessary. Attach one end of the second plastic hose to the upper

Drawing hot water from the solar water heater.

tube of the collector, but leave the other end free.

Fill the tank. When water flows from the open plastic hose, block the hose with your finger until water also flows from the top circulation tube of the tank. Then quickly slide the hose onto the tube and clamp it. This prevents air bubbles from being trapped in the lines. When the hot-water outlet overflows, close the tank valve.

Operation of the heater is simple. You will notice that the upper hose quickly gets hot, while the lower one stays relatively cool. Water is circulating now, and eventually all of it will be warmed by the coils. To draw hot water, open the valve at the bottom of the tank. Cold water comes in and forces hot water out of the top.

Water heater with reflector panel added.

To increase the heating capacity of your water heater, add a reflector panel that will also serve as a protective cover for the glass when you store the heater. One reflector, set at 30 degrees, will add 50 percent more heat. You can add four, if desired, and greatly increase the heating capacity. The water heater in the photograph uses one reflector, hinged at the bottom and secured at the top. A hole in the reflector fits over the 1/4 inch stud and is held in place with a nut when the heater is stored or being carried about.

Make the reflector(s) of 1/2 inch plywood, faced with aluminized Mylar, aluminum foil, or similar reflecting material. Attach to the heater with hinges, allowing one reflector to fold down on top of the other as shown on the drawing. Lengths of cord hold the reflector panels at the proper angle. As with the oven, when the reflector is closed it protects the glass cover of the heater.

You can use the heater in a more portable version, on a camping trip, for instance, where there is no water-pressure supply available. Simply close the inlet valve, fill the tank through the top, and let the water heat. To draw hot water, tip the tank and pour out water through the spout. Replace as needed.

You can very easily "scale up" the small heater presented here for use in a mountain cabin or a similar place that is not equipped with hot water. For example, a 50 gallon oil drum and a heat collector, 2 feet by 10 or 12 feet, mounted on the roof serves as a practical hot-water heater. Some swimming-pool owners have built their own solar pool heaters and saved the high monthly costs of gas-fired installations. A word of advice here: such an installation will require a very large area of heat collector (one-third or one-half that of the pool), properly oriented, and also a plumbing system, including a pump to circulate the water. A more challenging use of solar water heating is to heat a home or building instead of just its water supply!

Solar heater shown in drawing warms entire office building.
BRIDGERS & PAXTON

6

A Solar Still

NINETY YEARS AGO, mine owners in the high country of Chile were faced with the problem of providing water for their workers. The only available supply was unfit to drink, and so a means of purifying it had to be found. The solution was a sun-operated distilling plant in which a large area of glass-covered wooden frames

Solar still, located in the Andes Mountains in Chile in the 1870's, operated on the same principle we use in our project.
SOLAR ENERGY SOCIETY

evaporated the contaminated water, recondensed it, and thus produced 6,000 gallons of fresh water a day!

This Chilean solar still used no fuel or power except that from the sun's rays and provided pure water at a cost far lower than other means of distillation. But the method was forgotten for many years and fuel-operated stills were used whenever it was necessary to convert salty or otherwise undrinkable water.

Not until World War II were solar stills used again, except by experimenters. Fliers forced down at sea needed a source of drinking water until they could be rescued. Dr. Maria Telkes developed an inexpensive, lightweight plastic still that could be included even in one-man life rafts to provide a quart of fresh water a day.

Since that time, Dr. Telkes and other scientists have worked with solar stills of various sizes. The United States Department of the Interior is interested in the method, and suggestions have been made for large seacoast installations to purify salt water for drinking and irrigation. In some designs no pumps would be needed because the sea itself would fill the distillation tanks at high tide.

At present the cost of such a system is high, even considering that its operation would be cheaper than a fuel-run still. Engineers are hopeful, however, that improved methods and materials will make solar distillation economically competitive. Meanwhile, there are large experimental installations in Florida, Mexico,

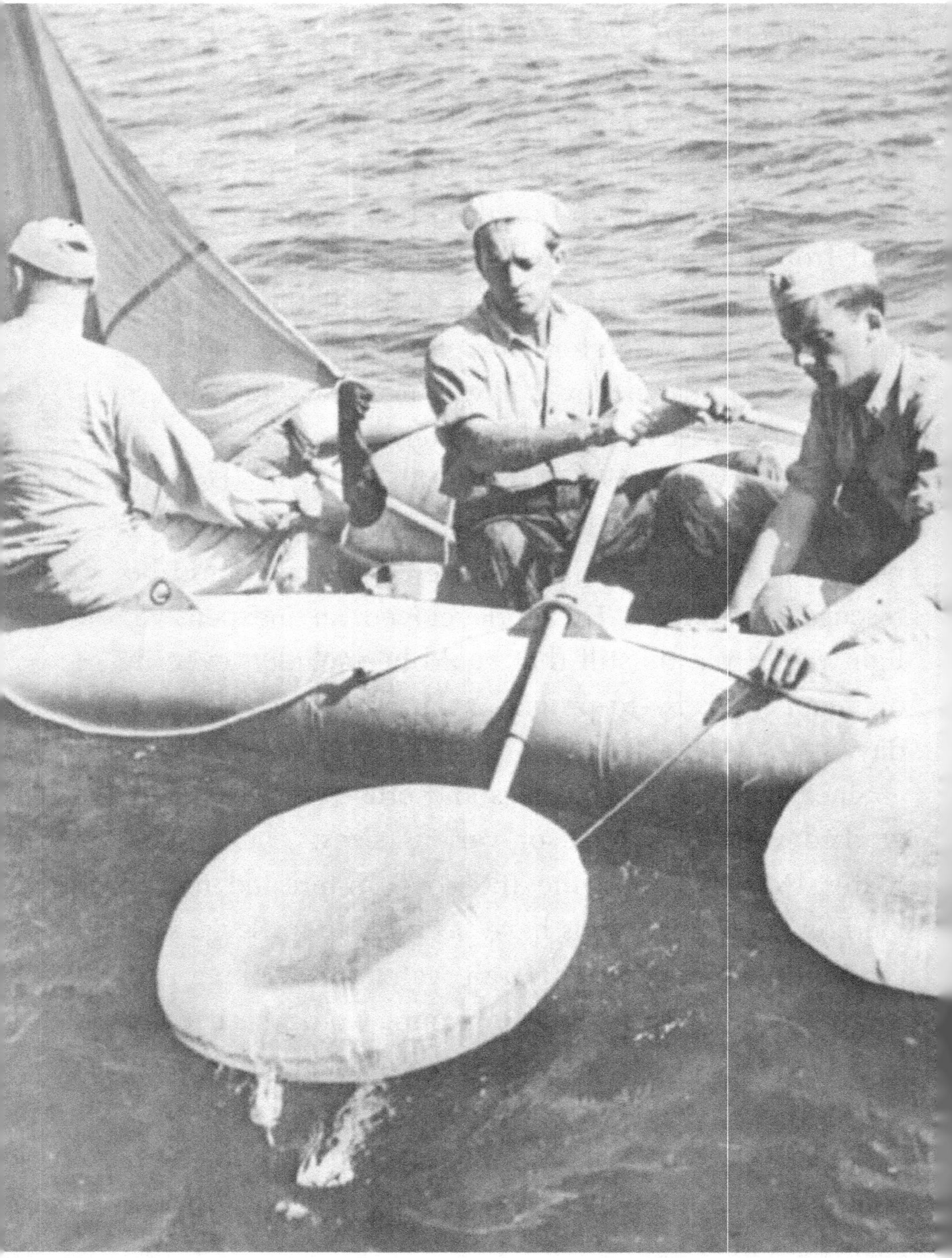

Plastic inflatable solar stills were developed for use in World War II by Dr. Maria Telkes.
SOLAR ENERGY SOCIETY

This solar still at Puerto Penasco, Mexico, was constructed by the University of Arizona and the University of Sonora. Its 12,000 square foot solar-heat collectors produce about 6,000 gallons of fresh water a day.
UNIVERSITY OF ARIZONA

Australia. On a number of Greek islands they are an important source of fresh water for the inhabitants.

The principle of the solar still is simple, and is observed on a grand scale in nature. Clouds are composed of water vapor evaporated from the surface of the sea or from damp ground and then condensed by cool temperatures high in the air until it becomes visible. In the process of evaporation, salt and other solids are left behind. Many readers will be familiar with the commercial harvesting of salt in shallow ponds, for this is one of the oldest of man's uses of solar energy. In solar stills the salt is a waste product, and we recover fresh water that is allowed to escape in the salt-pond operation. In the water heater, solar heat warmed the water in the coils. Since the water was enclosed in the coils and tank, little of it evaporated. However, in the still, the water is allowed to evaporate. Water vapor rises and contacts the underside of the glass cover. Here it condenses because the glass is cooler than the vapor. Gravity causes the droplets to run down the glass, along the bottom of the box, and out through the drain tube. Salt —or other pollutants—remain in the wick of the still or drain out in brine that is not evaporated.

The design has been proved by careful testing over a long period of time. Researchers have shown that it is possible to produce almost a quart and a half of water a day for each square foot of the collector's surface. Thus the still should have a maximum output of more than one gallon a day. Of course, this figure represents ideal conditions, but it will be interesting to compare your

This solar still produces fresh water from salty or otherwise contaminated water.

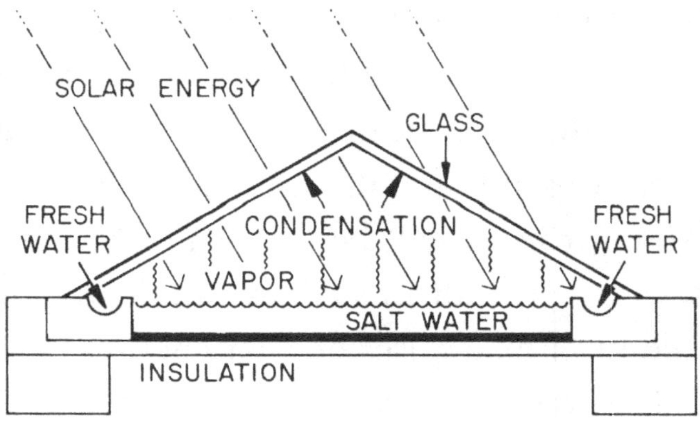

results with it. The still described here was tested in midwinter, and produced more than one quart of fresh water in six hours.

MATERIALS:

1 by 4 inch redwood board—8 linear feet
1/2 inch plywood—two pieces, 24 by 24 inches
Single-weight window glass—one piece, cut to measure
1/2 inch O. D. copper tubing—4-1/2 feet
1/2 inch, 90 degree copper elbow—one
1/2 inch copper pipe cap—one
Galvanized iron—one piece, 3 by 24 inches
1/4 inch I. D. copper tube—6 inches
Large tin can—one
1/4 by 1-1/2 inch wood screws—approximately thirty
Black terry cloth—approximately two yards (one large bath towel)
Carrying handle—one

To begin construction, first assemble all the needed materials (except the glass). Redwood is preferred where possible because it resists rotting, while other woods will deteriorate with constant exposure to water.

At the lumberyard where you buy the 8 foot board (1 by 4 inches) have a groove cut 1/8 inch wide to a depth of 3/8 inch. Locate the cut 1/2 inch from one edge. This is the slot for the glass window. Next cut the

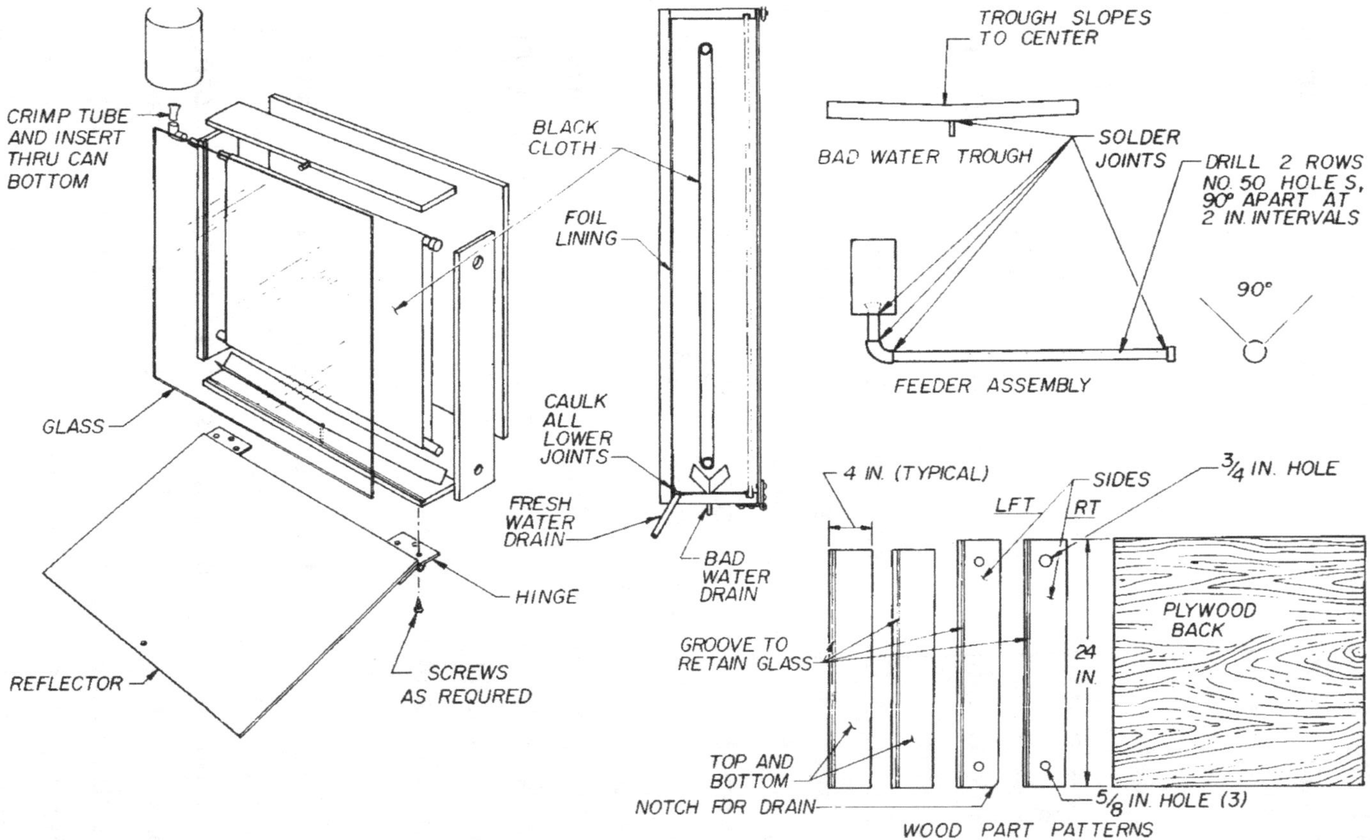
CRIMP TUBE AND INSERT THRU CAN BOTTOM
GLASS
REFLECTOR
BLACK CLOTH
FOIL LINING
CAULK ALL LOWER JOINTS
FRESH WATER DRAIN
BAD WATER DRAIN
HINGE
SCREWS AS REQURED
TROUGH SLOPES TO CENTER
BAD WATER TROUGH
SOLDER JOINTS
DRILL 2 ROWS NO. 50 HOLES, 90° APART AT 2 IN. INTERVALS
90°
FEEDER ASSEMBLY
4 IN. (TYPICAL)
SIDES
LFT
RT
3/4 IN. HOLE
PLYWOOD BACK
24 IN.
GROOVE TO RETAIN GLASS
TOP AND BOTTOM
NOTCH FOR DRAIN
5/8 IN. HOLE (3)
WOOD PART PATTERNS

SOLAR STILL

1 by 4 redwood sidepieces to the proper length, as shown in the drawing.

Drill two holes at each corner of the sidepieces and assemble with wood screws. The 1/2 inch plywood back may now be put in place and holes drilled for screws. Notice that no screws are put in at the corners, to prevent interference with those holding the sides together.

Mark the locations for the large holes that will receive the 1/2 inch tubing. Drill a 1/8 inch hole as a guide, and then drill three 5/8 inch holes and one 3/4 inch hole, as shown on page 77. Be sure the large hole is the proper size for the cap, which we'll solder to the end of the top tube.

The two drain holes can now be drilled, both of them 1/4 inch in diameter. One of these is located in the center of the bottom 1 by 4; the other is in the side 1 by 4, positioned at the vee formed by the bottom and plywood back. Study the drawing to be sure of locating this hole properly. The side drain can now be inserted.

This is a good time to calk the joints at the bottom and sides and apply several coats of sealer to the inside of the collector box. This is done to make it watertight, and so that the distilled water will run out the drain tube and not seep through the bottom of the box. Allow to dry thoroughly and check for leaks.

We know that any surface receiving heat will reradiate part of that heat. To prevent as much heat loss as possible, and also to present a smooth surface for condensation of water vapor, we line the inside of the box with aluminum foil. Running the foil in one piece across the plywood back and the bottom 1 by 4 will make an additional waterproof layer to help proper drainage. Notice that the foil extends into the bottom glass slot also.

Foil that is 24 inches wide will do the job in one piece. If it is necessary to use narrower foil, apply the section toward the drain tube first and lap the other section over it. Use glue, rubber cement, or airplane dope to apply the foil. Start at the top, carefully unrolling the foil for a smooth job. Let the edge of the foil extend a quarter inch past the edge of the glass slot. This excess will be forced into the slot later, when the glass is slid into place. Next line the inside surface of the other three 1 by 4's to complete the job.

Since some salty or otherwise undesirable water may not be evaporated before it reaches the bottom of the black towel wick, a vee-shaped trough is provided to catch this waste and prevent it from mixing with the distilled water at the bottom of the still. Made from galvanized iron, this 24 inch trough is a 90 degree angle with legs 1-1/2 inches wide. This can best be bent at the sheet-metal shop.

Drill a hole in the center of the angle, as shown, to receive the center 1/4 inch copper drain tube. Put the short length of tubing in place and solder securely. To keep the waste water from spilling out the ends of the trough, it is curved slightly by "crimping" the edges in several places.

Once the drain trough is completed, it may be installed by inserting the copper tube in the hole drilled for it. This should be a snug fit. Carefully punch through the aluminum foil first with a pencil, then press the trough down until it just touches the foil at the center.

We are now ready to start on the "wick" for the still. The 1/2 inch tubing used at the top and bottom to support the toweling is cut to size with a tubing cutter or fine hacksaw. Or ask your hardware dealer to do this for you, being sure to give him accurate dimensions.

In addition to holding the toweling in place, the top tube distributes water to the toweling. With a #50 drill, we drill two rows of holes at right angles to each other and spaced approximately 2 inches apart. These holes *do not* go through both walls of the tubing. See drawing for details.

The still uses a quart can as a reservoir. Cut the top from a large juice can and remove the paper. Next, cut a 4 inch length of 1/2 inch copper tubing and flatten one end in a vise or with a hammer. This tube will serve to meter the water into the long tube so that too much isn't fed to the toweling.

With a chisel, carefully cut a slit in the center of the can bottom. Force the metal inward until the flattened

Metal rods hold the towel "wick" in the foil-lined box of the solar still.

tube fits snugly with about 1/2 inch of it extending inside the can. Solder the tube in place, making sure it remains lined up.

Now slide the cap onto the far end of the drilled top tube and solder it in place. Push the tube through the holes in the box, being careful of the aluminum-foil lining. Prop up the box so that it is tilted back about 45 degrees from the vertical, and rotate the tube so that the two rows of #50 holes are properly positioned to feed water into the toweling when it is looped around the tube. The tilt of the still will vary with the position of the sun, so we are striking a happy medium in locating these holes.

With the tube in the right position, force the end cap into the 3/4 inch hole. This should be a snug fit and hold the tube in the proper place. Now solder the 1/2 inch elbow to the open end of the tube so that it points straight up. The free end of the tube soldered to the can is inserted in the elbow and soldered in place. Some water in the bottom of the can will prevent the joint at the can from melting while you work on the elbow.

Now attach a hinged 1 by 4 prop to the back of the box, using small wood screws as required. This supports the still and also adjusts the tilt to best face the sun. At this time also nail on the two legs, making sure to leave the proper one about 1/4 inch long so that water will run *toward* the fresh-water drain.

Smooth the ends of the bottom tube that will support the toweling and insert it in place. Now we are ready to sew the toweling together. The still uses a towel 24 inches wide, with the ends lapped and sewed in two places to make a loop. In measuring for this, make the loop slightly smaller than the distance between the tubes, because the towel will stretch somewhat when wet. Too much sag would cause it to touch the back of the box and thus contaminate the distilled water.

With the towel sewed together and dampened, slip the 1/2 inch tubes halfway out and start the towel loop onto the tubes inside the box. The bottom tube is slid back into place first; then the top tube is carefully raised into position and forced back into the 3/4 inch hole. Adjust the towel so that it covers the entire length of the tubes and is smooth. Do not let it touch the sides of the box.

At this point, remove the top 1 by 4 piece so that the dimensions for the glass can be taken. Slide a piece of cardboard into the grooves and trim to fit. Have your glass supplier cut the glass to measure. The grade of window glass known as "water-white" allows more of the sun's rays to pass than ordinary glass does and is therefore more efficient for our purpose. However, if it is not available, standard single-strength glass will do.

Slide the glass into position. Make sure it is clean, particularly on the inner surface, which will not be easy to reach when the still is assembled. Detergents may interfere with proper forming of droplets on the glass, so use plain water for cleaning.

Handle the glass very carefully to avoid cutting yourself on the sharp edges. With the still at a 45-degree angle, start the glass into the slots and ease it downward. When it contacts the aluminum foil at the bottom groove, it should force the foil neatly into place. Replace the top piece, tighten the screws, and the still is ready for operation.

Distillation is easy with the solar still. First, orient the still so that the sun's rays strike it as close to a right angle as possible. Mount the still on a level surface so

Operation of the solar still. Salt water poured in at top condenses on glass, as shown, and drains out at bottom.

that the uneven legs will give it the proper slant for draining. If the reservoir can is not vertical, carefully twist it until it is. The towel should be snug enough to hold it, and the force fit of the end cap in the 3/4 inch hole helps in this respect.

Now fill the reservoir, and to make sure that the still really works, add some salt to the water! After a few minutes the towel will begin to receive water from the distribution tube. If it doesn't or if the flow is too slow, very carefully open up the flattened tube in the bottom of the can with the point of an ice pick. Be cautious about this, however, as the water should not flow into the towel faster than it can be evaporated.

Heat from the sun is trapped in the box, and the black towel absorbs this. Water is therefore evaporated from the towel much faster than it would be normally. The vapor, free of all solids, recondenses on the smooth surface of the glass and the foil on the back and sides. You'll see this a few minutes after you put in the water; first the glass steams up, then droplets form and run down to the bottom.

Don't expect fresh water to gush from the still like a Niagara, but put a can or bottle under the drain to catch the distilled water. On a sunny day the still will begin producing soon after you set it in operation and will drip water steadily into the container. To guard against evaporation of this distilled water, some experimenters run a tube from the drain to a corked bottle.

At first you will probably reposition the still every

half hour or so for greatest efficiency. When the novelty wears off and you want to leave the still for steady operation, decide on a compromise tilt—your latitude plus 10 degrees or so—and aim the collector south (unless you live below the equator, of course).

To boost the capacity of your still use the reflector method described before. Make up one or more reflectors as with the water heater and attach them to the body of the still.

Addition of reflector panel increases yield of the solar still.

These three men are installing earth-water stills at Mount Mihara on Oshima Island.
SOLAR ENERGY SOCIETY

THE SURVIVAL STILL

There is another kind of solar still, which is even more remarkable than those previously described. Called a "survival still" or "earth-water collector," this type draws moisture out of the ground, evaporates it, condenses it on a glass or plastic cover, and collects the fresh water in a container.

of producing fresh water on islands of the Pacific where none was available much of the time. Experiments were made with earth-water stills but no practical use was made of them. In the 1960's, however, a Japanese scientist named Kobayashi improved on the wartime experiments and designed a small portable still that produced a quart of water a day from the soil outside the Nippon Electric plant, where he was employed. Further tests showed that even in hot, dry deserts, the still produced some water.

In 1965, Dr. Ray Jackson and Dr. Cornelius van

Design of the survival still. Cactus or other water-yielding vegetation may be added under the plastic to produce more water. The plastic drinking tube permits water collection without dismantling the still.

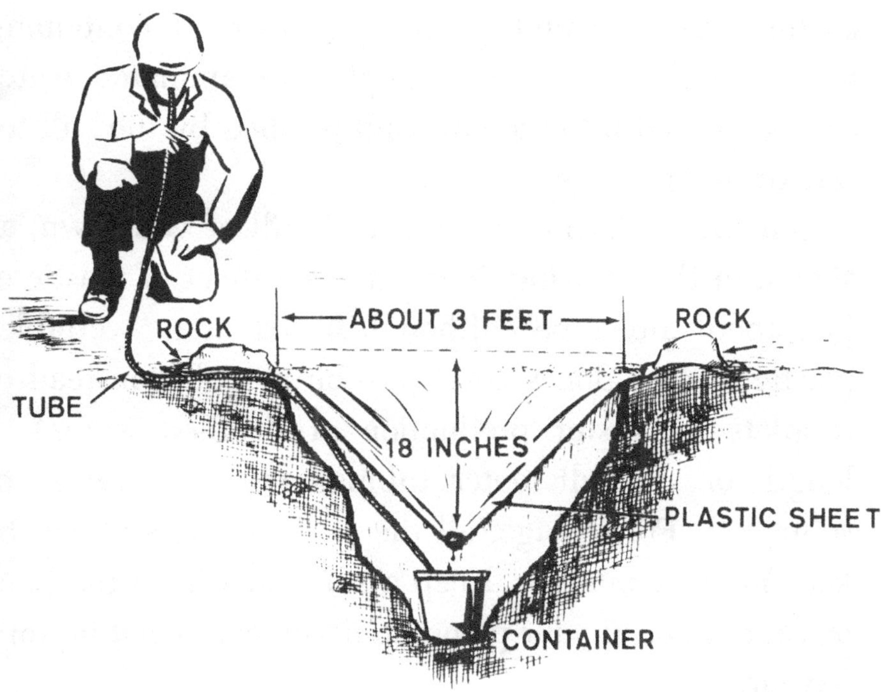

Bavel of the U.S. Department of Agriculture Water Conservation Laboratories in Tempe, Arizona, developed a much simpler survival still independently of the Japanese research. The Japanese had simply scraped off a bit of the top soil and placed their still on the ground. The Tempe scientists dug a conical hole in the ground and covered it with a sheet of plastic. Dirt from the hole weighted the edges of the plastic, and a rock or other weight formed it into a cone reaching down into the depression. The results were spectacular, and some test stills produced a quart of water or more per day for as long as a month at the same hole!

Refinements to the survival still include addition of a plastic tube to allow someone to drink the water without disturbing the still itself. It is also helpful to add cactus, other vegetation, or any material containing water, to the hole. For example, contaminated water can be poured into the hole and purified by the still for safe drinking.

You can easily make a survival still of your own, as shown in the drawing. Needed are a sheet of plastic of just about any type (although Tedlar is a "wettable" plastic and produces condensation in a film instead of droplets for faster production of distilled water), a length of small-diameter tubing, and a container of some sort. For storage, all the parts of the still can be kept in the container. Keep a survival still in the back of your car or in your plane, so if you ever need it you'll have it.

Drinking from a survival still. These devices produce as much as a quart of water a day.
U. S. DEPARTMENT OF AGRICULTURE

7

Radio

OUR PROJECTS so far have used the heat of the sun directly, and all have been quite easy to understand. In this chapter we take up a sun-charged power supply that makes use of the *photovoltaic* process, discovered by Antoine Becquerel in 1839. The conversion of light directly into electricity—with no intervening steps—is as fascinating as it is difficult to understand.

It was a long time from the discovery of the principle of photovoltaic conversion to its practical application. At first it seemed no more than a scientific curiosity because of its low efficiency. Early photocells used the element *selenium,* which derives its name from the Greek word for moon. This is an odd name for a material now used in so-called *sun* batteries. The selenium cell was first used in photographers' exposure meters, electric-eye devices, and the like. Present-day solar cells of selenium are inexpensive; their only drawback is a relatively low efficiency of conversion—less than 1 percent.

Pioneer experimenters with transistors used the element silicon, then switched to germanium. The unusual properties of silicon were remembered later, however, when efforts were made to develop a more efficient photovoltaic converter. Bell Telephone Laboratories produced solar cells with an efficiency of more than 6 percent. Bell then proved the usefulness of solar cells by powering rural telephone lines with sunlight. Then space applications took over and the solar battery became famous.

In a solar battery bundles of energy, called photons, penetrate the selenium or silicon and cause the movement of electrons through the material. If a wire is connected between the positive and negative terminals of the battery, current flows. This flow of current is similar to that in a dry cell, with one important difference. Nothing is added or taken away from the solar cell; it will produce current indefinitely instead of wearing out, as the dry cell will.

What this means may best be understood in terms of our space program. Conventional batteries are bound to wear out. But, barring damage from cosmic rays or some unforeseen occurrence, solar batteries will just keep operating—perhaps forever! In addition to their use in satellites, solar cells are used for radio transmission in remote locations and as warning beacons. Clocks and radio receivers operate twenty-four hours a

Artist's concept of a remote weather-reporting station powered by a large solar panel.
HOFFMAN ELECTRONICS CORP.

day, using solar cells coupled to storage batteries. The self-charging flashlight is another successful use of solar cells.

Solar-cell panels of many square feet are available commercially for tasks that justify the high initial cost. Solar cells are not cheap, even though silicon is one of the most plentiful elements. To be suitable for solar cells, the silicon must be ultrapure, and the raw material to make a sun battery is costly. The cell-manufacturing process is expensive too. It is helpful to remember, however, that the first cost of a solar battery is total cost.

The recent introduction of transistors and improved solar converters has made possible radio reception, and even transmission, with no power other than sunlight. In 1955, a message was sent from a sun-powered transmitter and picked up on a sun-powered receiver operated by radio hams. Our armed services and NASA have developed walkie-talkies, helmet radios, and emergency equipment powered by solar energy.

More recently, sun-powered transmitters have been sent aloft in satellites. Sun batteries also power remote radio installations and harbor warning beacons by charging storage batteries for round-the-clock operation. The advantages are obvious. No power lines are necessary, and solar batteries last indefinitely.

Now let's build a radio powered entirely by sunlight, or artificial light for indoor and nighttime use. Our radio makes use of a selenium cell that can be pur-

A possible use for the solar-powered radio. This experimental transmitter-receiver was powered by solar batteries mounted on soldier's helmet.
U. S. ARMY PHOTOGRAPH

chased for as little as a dollar and a half or a silicon cell that costs slightly more. The practical application of solar-powered radio has passed far beyond the simple receiver described here; however, the easy-to-make little radio nicely demonstrates the use of the solar battery in electronics. The interested reader will find a large amount of material on more advanced projects in handbooks and in mechanical and electronics magazines.

MATERIALS:

B2M solar battery (or substitute)—one
audio transistor—one
IN34 diode—one
365 tuning condenser—one
Antenna coil—broadcast, one
Headphone—2,000 ohms or more, one
Plastic box—one
Phone jacks—two
Binding posts—two
Knob for condenser—one
Knob for antenna coil—one
Terminal board—one
Rubber feet—four
No. 24 hook-up wire, antenna wire, and 6-32 screws, nuts, and washers as needed

Most of the parts for the radio may be purchased at your local radio wholesale house or from a mail-order firm. You will be able to buy from the wholesaler if you tell him you are an experimenter and describe the project. The plastic case—a sandwich box or similar container—can be bought at the dime store. Prices will vary somewhat, but will be between five and eight dollars for all the parts, including the earphone. In case the exact parts specified in the list of materials are not available, the clerk will be able to provide equivalent items.

No attempt has been made to miniaturize the assembly of the radio, and the beginner will find it easy to put together. The various parts are spaced for easy connection of wires and ample room for the soldering iron. If desired, the radio may be built into a box one-quarter the size of that shown on page 98. This chapter includes such a project but it is for those with some previous experience with radio work. Whichever plan of attack you decide upon, have all the parts on hand before you start work. You'll find that you can complete the radio in an evening, and start enjoying it immediately—using a lamp to substitute for the sun.

The first task is to drill all the mounting holes in the plastic box which will house the radio. Check the diameters of the phone jacks, binding posts, antenna coil, and condenser at the points where they go through the plastic, and drill holes of the proper size. For the larger holes it is a good idea to drill a pilot hole first and then enlarge it to the correct size. Be careful not to crack the plastic. If you do not have a large enough drill, use a round file to do the job.

Because the condenser shaft must protrude through the plastic cover, drill this hole first. Locate it according to the dimensions on the drawing, unless the condenser you are using is of much different size and shape. Place the condenser in the box and put the lid on it. Now turn the box over carefully, keeping the condenser properly located, and drill through the plastic and into the mounting holes in the condenser frame.

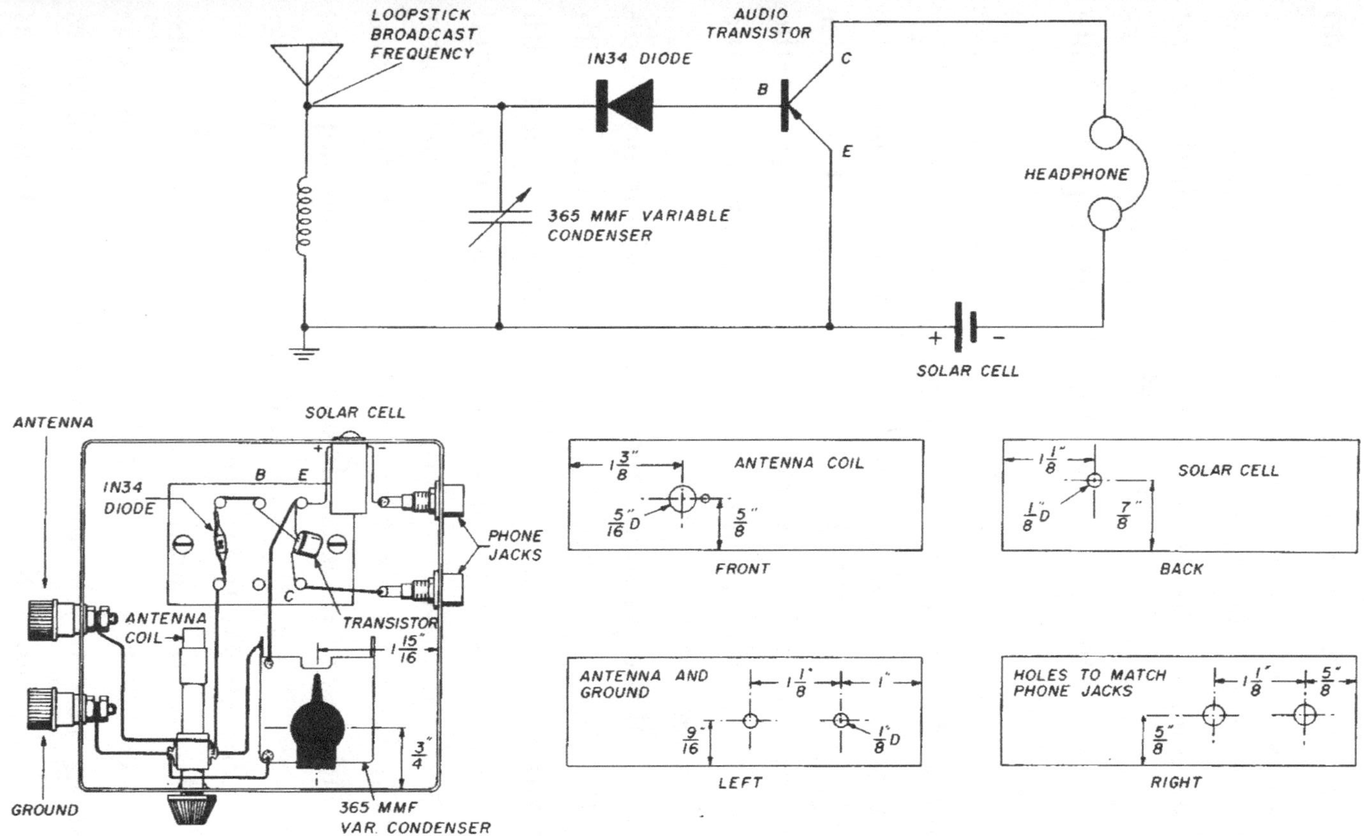

SUN-POWERED RADIO

This procedure ensures that all the holes will line up when it is time to assemble the radio.

Notice that the small hole drilled near the edge of the antenna-coil mounting hole must match the metal fitting on the coil. Use the metal mounting bracket included with the coil as a template to properly locate the small hole.

After all the holes are drilled, including those for the terminal board, clean up rough edges, remove chips from the box, and assemble the parts. Tighten the phone jacks and binding posts snugly, and use fiber washers with the 6-32 screws and nuts to protect the plastic. Snap the antenna coil into place and mount the solar battery with the glossy surface up.

With all the parts installed, do the wiring next. Follow the drawings carefully if you are not familiar with radio theory. The diode will probably be labeled with an arrow, like the one shown on the drawing, or coded in some other way to indicate direction. The center lead on the transistor is B, as shown in the drawing, and the lead closer to it is E. Make sure of the hookup of the diode and transistor and also that the sun battery's black wire goes to the phone jack.

Soldering the connections will give best results. However, since it is easy to damage transistors by overheating, you may want merely to wrap the lead wires around the posts of the terminal board. While not as good an electrical connection, this method does work and was used on the radio in the photograph. If you do

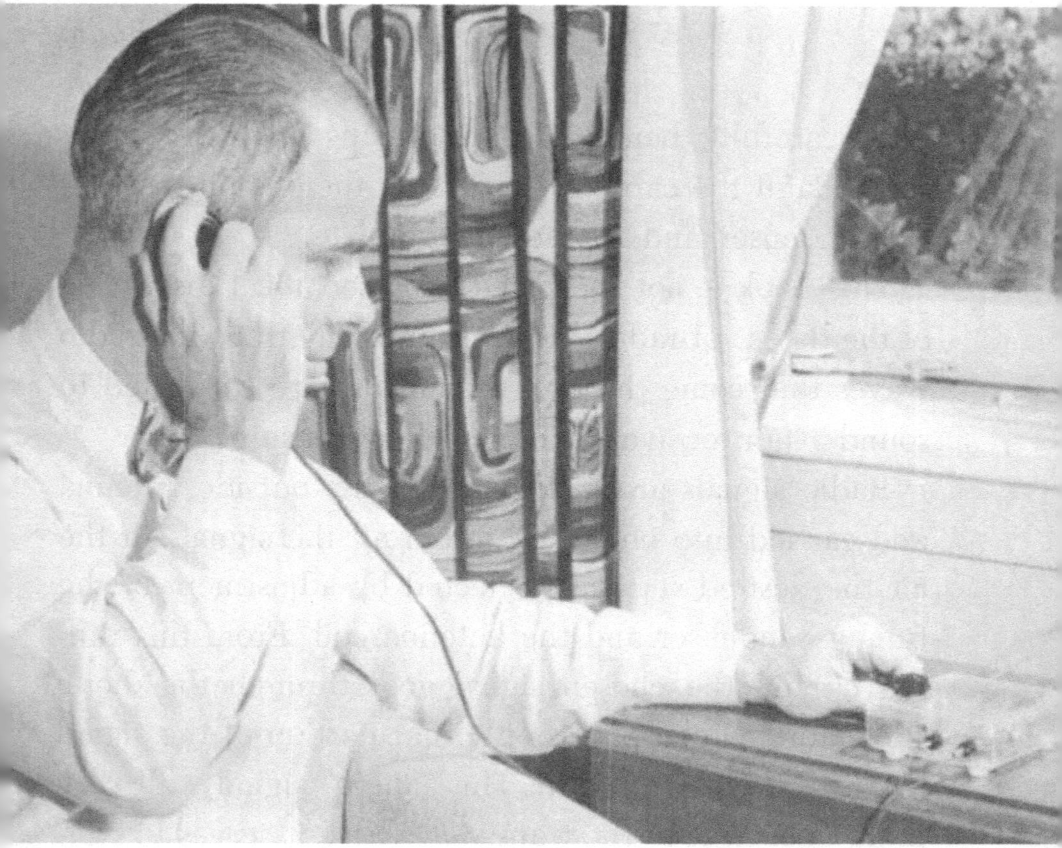

The solar radio works indoors, too, using light from a window or lamp.

solder the transistor in place, pinch the transistor lead between the solder joint and the transistor with a pair of long-nosed pliers during the operation, to draw off surplus heat.

Now wipe up any excess soldering paste; check all joints and mounting screws for tightness. If everything is in order, put on the plastic top and mount the control knobs. If you were unable to find a knob small enough for the antenna-coil adjusting screw, use the cap from a toothpaste tube as shown in the photograph. Drill a hole slightly smaller than the screw and twist the cap onto the shaft. To make the adjusting screw easier to turn, carefully bend the spring clips away from the shaft. Paint the cap black to match the one on the tuning condenser, and the radio is complete.

This book is not the place for a technical discussion of the theory of radio, but we can briefly trace the radio waves that come through the air and are converted to sound in our receiver with the help of sunshine.

Radio signals are picked up by the outside antenna and carried into our radio. From all the signals in the air the desired station is selected by adjustment of the tuning condenser and the antenna coil. From this tuning circuit, the chosen signal goes through the diode "detector," which passes only that portion of the signal that we are interested in. This "audio" signal, still very faint, goes next to the transistor where it serves to trigger proportional, but much greater, amounts of power from the sun battery. This amplified audio signal activates the earphone and makes the music or the news broadcast we hear.

So much for theory; now to see how well our solar radio works. Attach an outside antenna to the proper binding post and a ground wire to the other post. The longer the antenna, the better, although the radio shown performed well with about 10 feet of wire and brought in several stations. It did a better job when connected to the TV antenna.

If you live many miles from the nearest radio station, you will have to use a long antenna. In extreme cases it may be necessary to increase the power of the radio by adding one or more solar batteries, connected to the original in series—that is, red wire soldered to black. This increases the voltage of the power supply.

In most cases the radio will bring in several stations nicely in bright sunlight. Use the tuning condenser to select your program and sharpen up the adjustment with the tuning coil. To fancy up the radio, add a broadcast dial, with appropriate frequency numbers drawn in as they are on a larger radio. You've probably noticed that there is no on–off switch, but don't worry about leaving this radio on all the time. There's nothing to wear out, and the power is free!

At night or on very cloudy days, you can tune in your favorite programs by placing the radio under a lamp. Don't place the solar battery so near the bulb that it becomes heated, as this may damage it. This should not be a major problem; satisfactory operation can be obtained at a safe distance from the lamp.

That's the solar radio. Good luck in building and using it; with care your set should give a lifetime of service. If the electronic side of solar energy appeals to you, there are many interesting projects you can work up: a code practice oscillator, for example, a continuous-wave (code) transmitter, or even a portable receiver that requires no external antenna.

Solar mite

The solar radio is a challenge for the miniaturizer. So

a compact, "space-age" version of the basic design is given next. As you will see, the circuit is the same as the one used in the radio just described, but smaller components are used in some cases and the whole assembly is shoehorned into a tiny plastic box. Even though it is much smaller, the Solar Mite is a better performer since it packs its own collapsible antenna. The tuning dial is more attractive and has its own calibration.

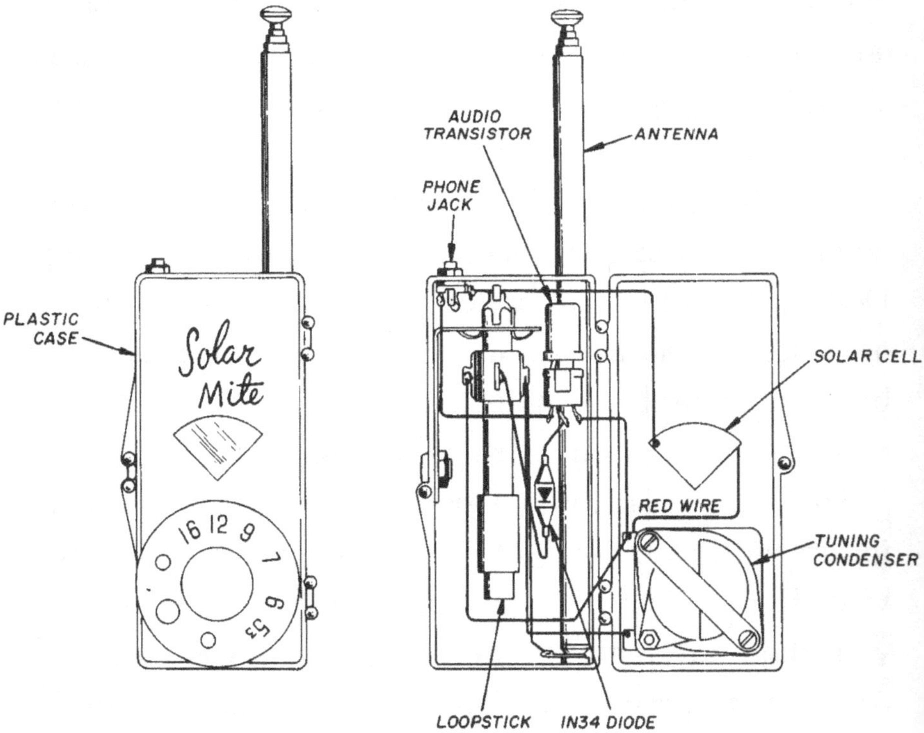

MINIATURIZED RADIO

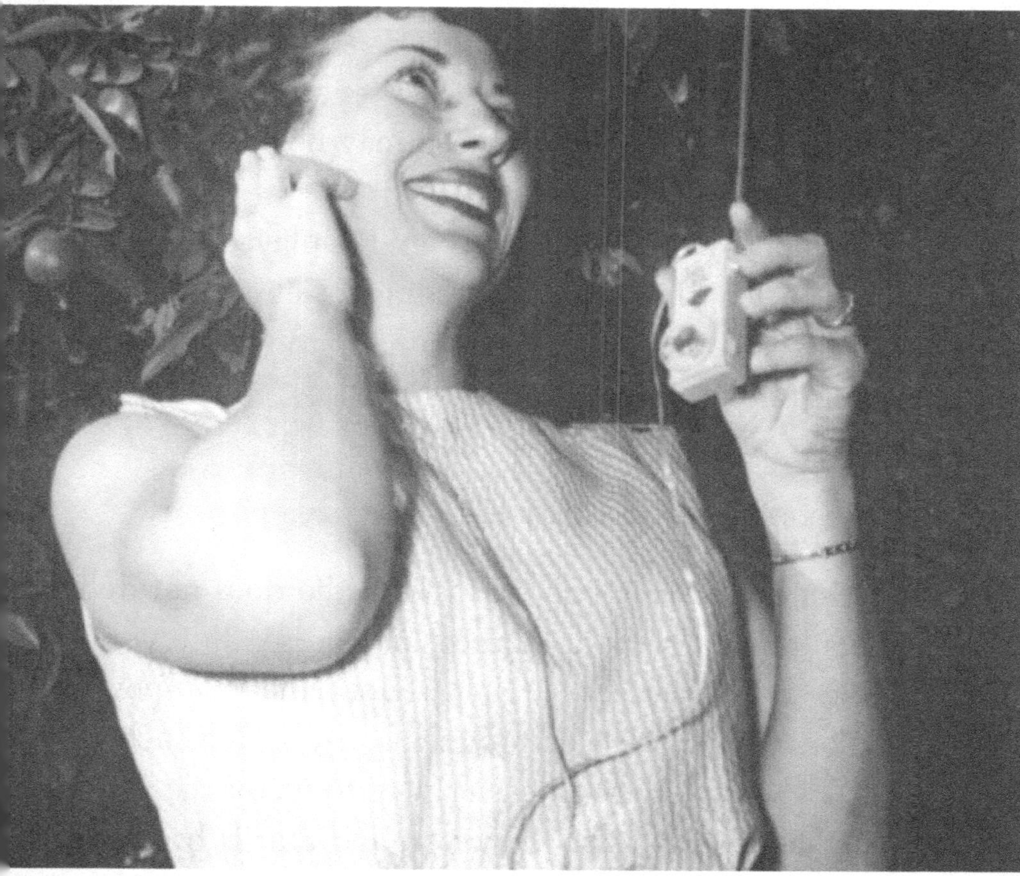

Solar MITE *with antenna extended for outdoor reception.*

If your local radio supplier can't provide the necessary parts, check mail-order firms like Lafayette, Olson, Radio Shack, or others. The radio in the photograph is sprayed with orange Day-glo paint and is an eye-catcher. Don't paint the box cover, but cut a piece of orange paper to fit inside it. Letter as desired, and cut out for the solar cell. See page 103. The point of the cell acts as the station selector. The quarter-round silicon cell used in this project was broken from a round cell. Place cell on a magazine, tap lightly in center with a center punch and—presto! Four perfect quarter-sections, each big enough for a solar radio. Of course, a standard 1 by 2 centimeter cell may be used. Just cut a hole in orange paper to fit.

Indoors, the Solar Mite can be played beneath a lamp. A length of wire with alligator clips at each end can be used to attach the antenna to the lamp or to the phone for better reception.

Convert to solar power!

A one-solar-cell, one-transistor radio does not deliver the ultimate in performance, of course. But you can build more complex, higher-performance sun-powered radios. Check the electronics magazines for such projects. There is an even easier way to enjoy the fun and benefits of solar radio, however, and that is by converting a pocket transistor radio to solar power.

A single selenium or silicon solar cell will operate only a small earphone-type radio. Five silicon cells soldered together will provide ample power for the popular pocket-sized radios, and will drive the loudspeaker to plenty of listening volume. You can put together one of these solar "shingles" yourself, or buy it ready-made and covered with protective plastic. Some companies provide the solar pack with attaching wires and a mounting device.

If you plan to use a five-cell battery you will have about 2 volts to work with. This will operate a radio normally using two pencells, or 3-volt supply. For a radio using a 4.5 or 9 volt battery you will need more

ALL-SOLAR RADIO

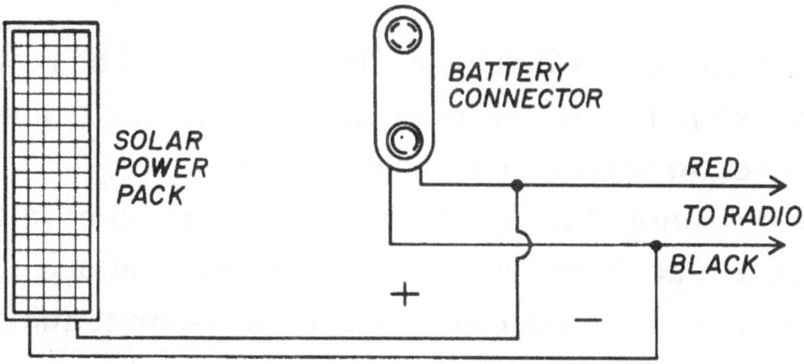

SOLAR OR DRY CELL

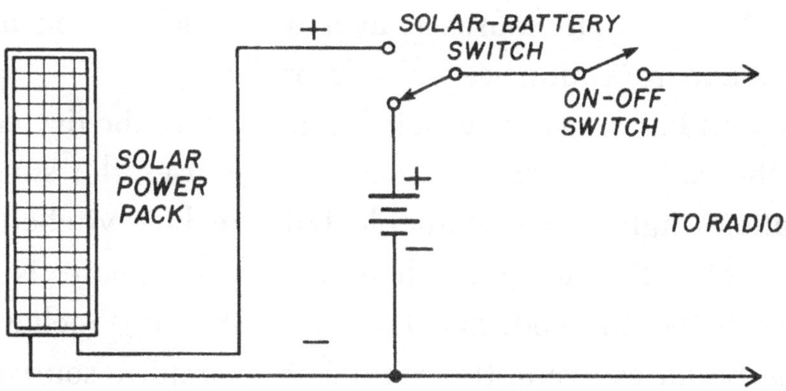

SOLAR-RECHARGEABLE BATTERY

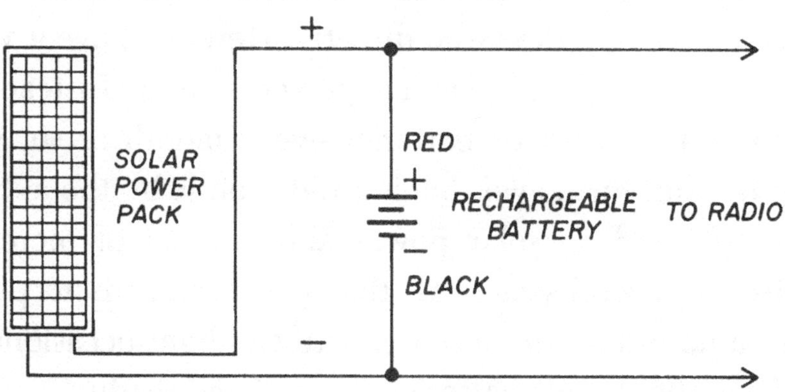

SOLAR CONVERSIONS

cells. Each solar cell delivers about 1/2 volt. The power packs offered by suppliers are available in 4.5, 6, and 9 volt sizes, so get one to match your radio.

In the simplest conversion to solar power, remove the conventional batteries and permanently attach the leads from the solar cell to the battery-box terminals. Now you will have a radio that will play when it is in sunshine or under artificial light of enough intensity. This is fine except during nighttime or cloudy weather, or when no light source is available indoors. You may want a more "sophisticated" solar radio.

By adding a two-way switch and leaving the dry cells in the radio you can have a set that uses the sun or artificial light when available but can be switched to conventional batteries when necessary. There is an even better method, and that is substituting rechargeable batteries inside the radio as the standby source of power. In this circuit, as the drawing shows, the solar battery charges the mercury batteries in the radio when it is not playing the radio directly. Here we have a very convenient and long-lasting power source. Instead of buying a new set of batteries every month or so, you tap the sun for power. Such a radio built by the author has operated on solar power and one set of mercury cells for several years. All that is necessary is to place the radio in the sun or under artificial light occasionally to keep the storage batteries charged and ready.

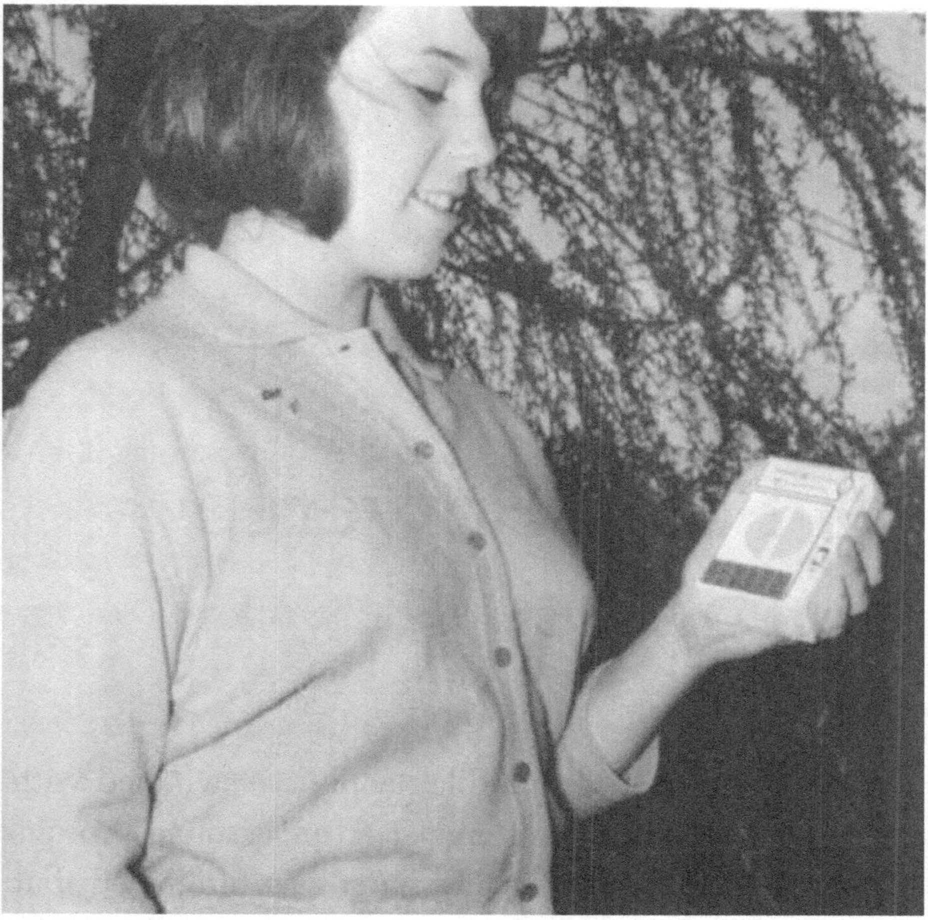

This transistor radio has been converted to operate on either conventional batteries or solar power, at the flip of a switch.

8

A Solar-Powered Airplane

From time to time gullible people are swindled with engines that run on water instead of gasoline. An airplane that flies on sunshine sounds as farfetched, but the model shown in this chapter does just that. It is presented last, as the most advanced and intriguing project demonstrating the uses of solar energy.

Indirectly, glider pilots have been flying on solar energy for many decades. This is so because the sun heats the earth and produces helpful "thermal" currents of air that rise and carry soaring craft up with them. Our project is not a glider, however, but a powered craft. It mounts a tiny electric motor driven by an array of solar cells of a special kind. With the sun overhead, enough electricity is generated to drive the propeller fast enough to fly the plane. The fuel for this amazing model airplane is weightless—unless you can

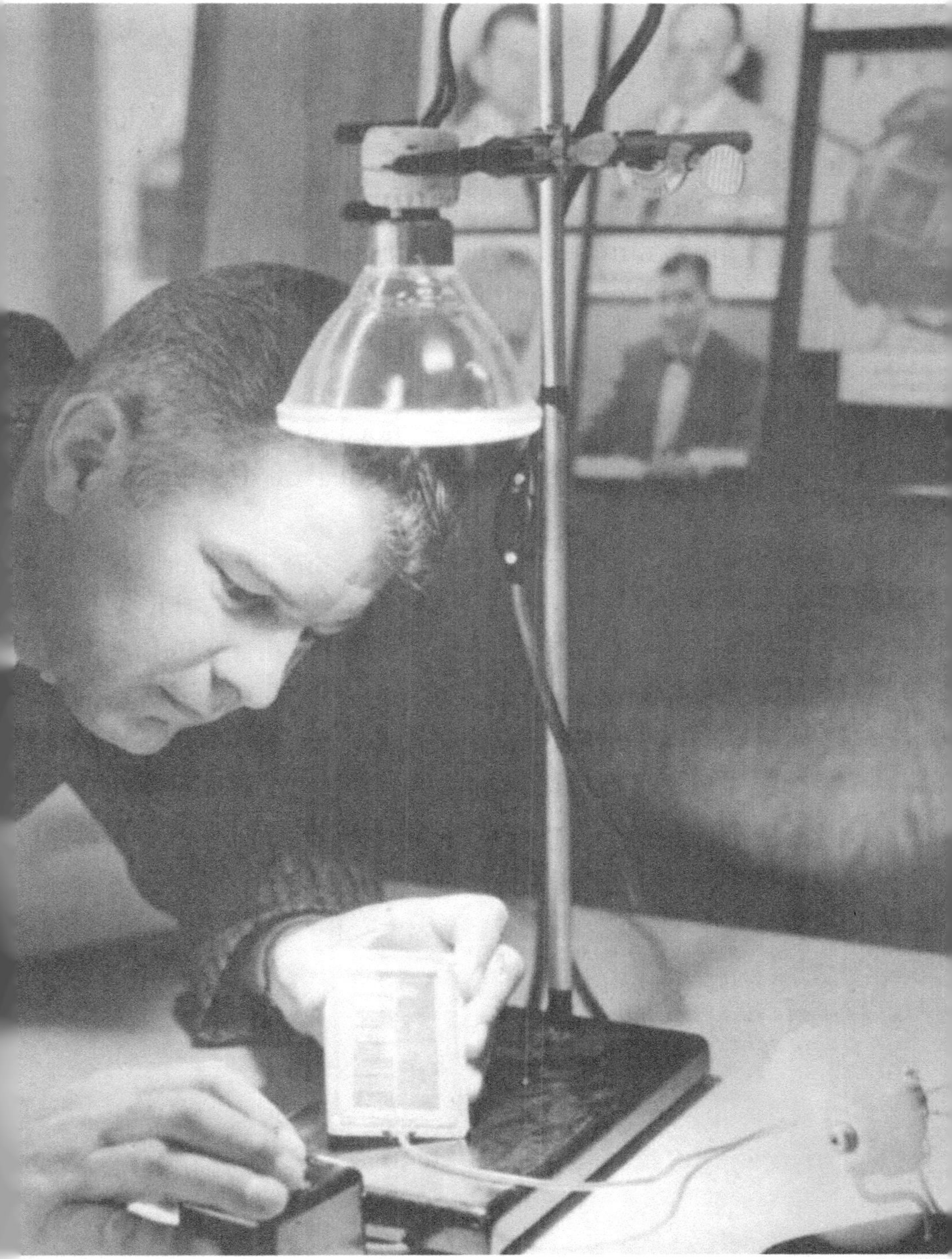

Germ of the solar-powered airplane. Engineer demonstrates conversion of light energy to power to drive a small electric motor and propeller.

U. S. ARMY PHOTOGRAPH

weigh a sunbeam—and the craft will fly as long as there is sufficient sunlight.

Soon after solar batteries were introduced they were used in a number of demonstrations. Some furnished power for electric motors driving the propellers of scale-model airplanes. While this made an eye-catching display, solar-battery manufacturers were quick to point out that their products were not being suggested to operate airliners! However, the idea of a solar-powered model airplane was inevitable.

The airplane presented here is one of a series of designs built during a period of several years. Overly optimistic about the amount of power in sunlight, I first built a conventional model plane with a wing area of some 200 square inches, and an over-all weight of close to 5 ounces. This sounds quite light, but it might as well have been 5 pounds, even with a total of 100 solar cells mounted atop the fuselage!

The successful version of the solar plane is simple in design and light in weight, to fly on sunbeams.

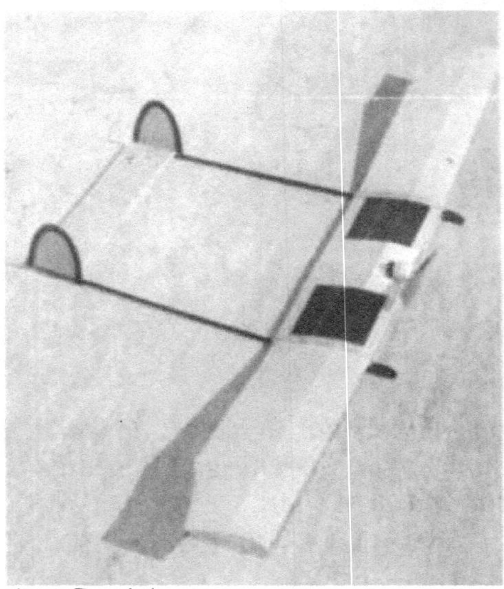

Sadder and smarter, I shrank the size of the craft, eliminated the heavy fuselage, and shaved weight in every other way I could think of. Each new design weighed less and crept nearer to the goal of sustained flight on sun power. But only when a new kind of solar cell (weighing about one-third as much as conventional silicon cells) became available was success achieved. Conventional solar cells are made from slices of high-purity silicon, "grown" in single crystal form in a molten furnace. There is also a solar cell made from cadmium sulfide, using the "thin-film" technique. A thin-film cell is deposited as a vapor on a suitable "substrate." Produced by Clevite Corporation, cadmium-sulfide solar cells are marketed in sizes up to 3 inches square, compared with the standard silicon cell only 1 by 2 centimeters. Furthermore, the cadmium-sulfide cell is only a few thousandths of an inch thick, flexible enough to be formed over curved surfaces, and only about one-third the weight of silicon cells of the same power output. Presently on the market are cells of 6 percent efficiency, only about half the maximum of silicon cells. However, the other advantages outweigh this shortcoming in many applications.

The 3-inch Clevite cells each produce about 300 milliwatts of electric power at a little less than 1/2 volt. Sliced into thirds and wired in series, two of them yield about 2 volts and drive a tiny electric motor and its geared-down propeller. To boost the craft's power, "solar-chargers" are added. These foil reflector panels—

similar to those we used on previous projects—concentrate more sunlight on the solar cells and up the power output considerably.

For the experienced model builder the solar plane will be easy to construct. Beginners will have to work a bit harder but the results far outweigh the work entailed in building the project.

MATERIALS:

3 inch cadmium sulfide solar cells (Clevite)—two
Micro-Mo motor 05L—one
four-inch propeller, plastic or wood—one

balsa wood:

1/8 by 1/4 by 18—two pieces
1/16 by 1/16 by 36—two pieces
1/16 by 1/8 by 36—two pieces
1/32 by 3 by 18—two pieces
Copper wire (single strand), 24 inches
Tissue paper—one sheet
Glue
Sheet metal for switch
Paper cement
Rubber bands
Timer fuse

The plane is built of balsa wood for light weight. The wood strips and sheets chosen should be a compromise

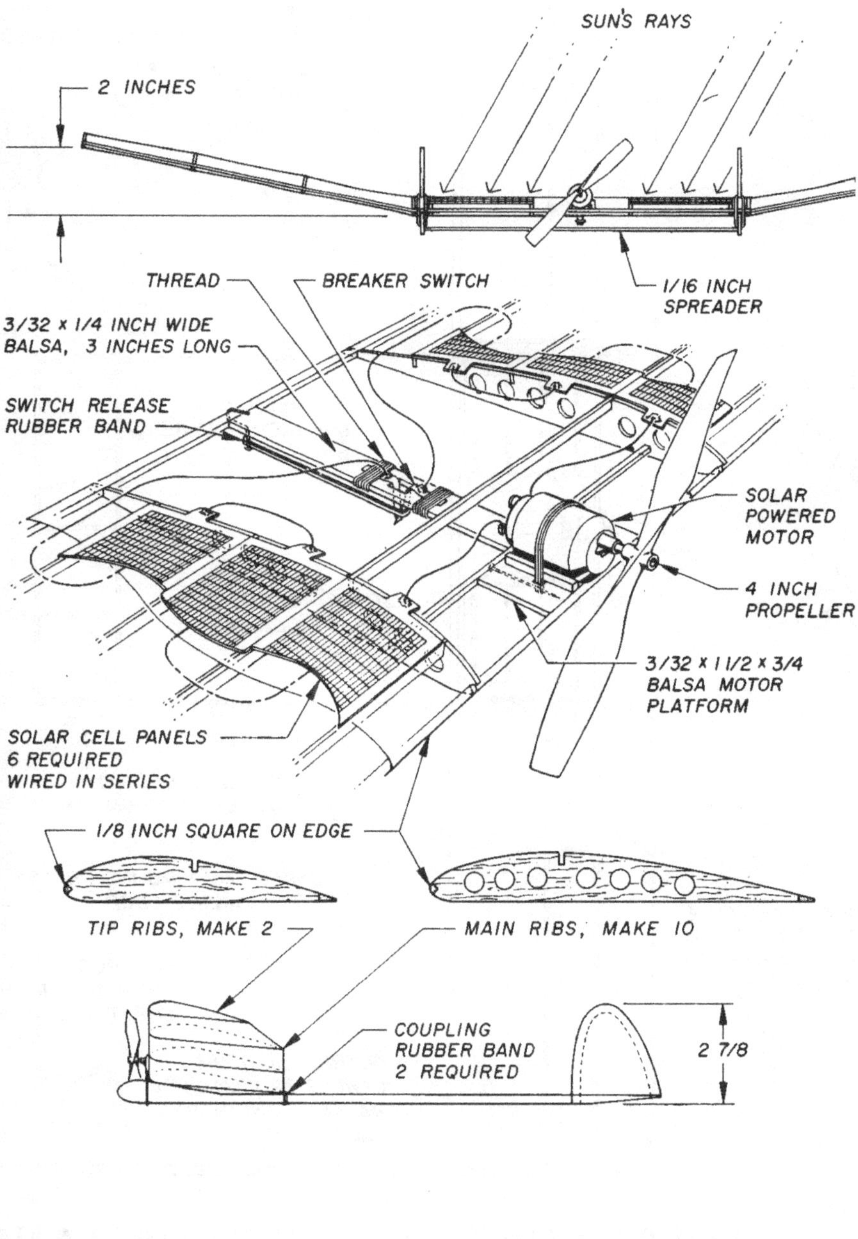
SUN'S RAYS
2 INCHES
1/16 INCH SPREADER
THREAD
BREAKER SWITCH
3/32 x 1/4 INCH WIDE BALSA, 3 INCHES LONG
SWITCH RELEASE RUBBER BAND
SOLAR POWERED MOTOR
4 INCH PROPELLER
3/32 x 1 1/2 x 3/4 BALSA MOTOR PLATFORM
SOLAR CELL PANELS 6 REQUIRED WIRED IN SERIES
1/8 INCH SQUARE ON EDGE
TIP RIBS, MAKE 2
MAIN RIBS, MAKE 10
COUPLING RUBBER BAND 2 REQUIRED
2 7/8

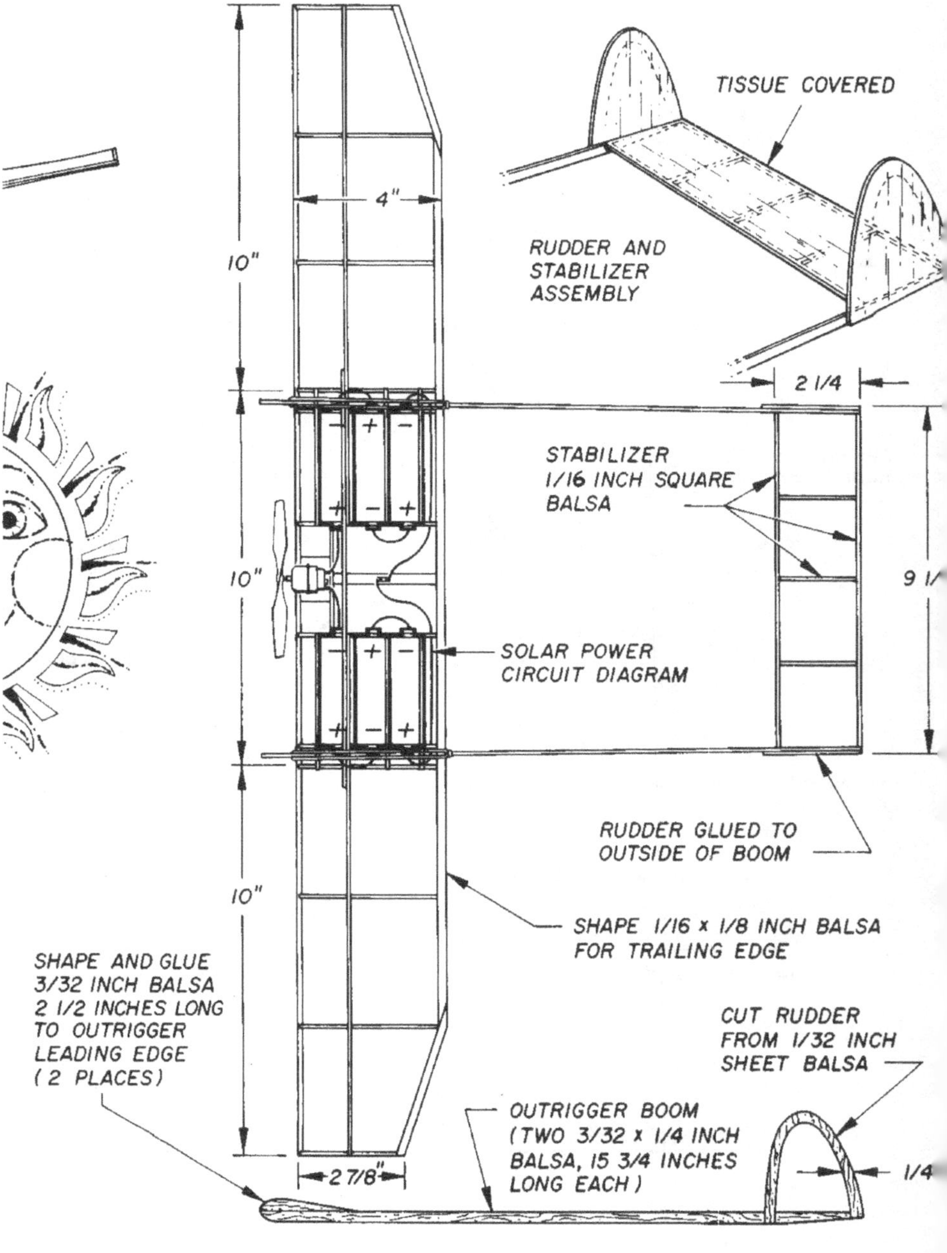

SOLAR POWERED MODEL AIRPLANE

between weight and strength. The main wing spars should be hard and strong since they are stressed more heavily than the other members. The leading and trailing edges of the wings, the booms, and the tail surfaces should be of medium-hard stock. The wing ribs can be made of soft balsa. The ribs are further lightened with holes made by a paper punch.

Build the wing first. It is constructed in three separate panels and then the outer panels are glued to the center with the wingtips raised for the proper dihedral angle. While the glue on the wing joints is drying, make the stabilizer and the twin rudders, as well as the booms that attach the tail surfaces to the wing. Assemble the stabilizer to the booms and allow glue to dry. Carefully sandpaper the leading and trailing edges to a rounded shape; then cover the stabilizer with paper on the top side only. When this is done cement the rudders in place. Allow the glue to dry and then cover the rudders with paper on the outside only.

Add another coat of glue to the wing joints. Round off the leading edge and shape the trailing edge to a knife edge as shown. Sandpaper smoothly. The wing is now ready for mounting of the solar cells.

Trim off excess metal from solar-cell terminals, leaving only a small tab. The six cells are now glued to the ribs and strips. See pages 114–15. The polarity of the cells is reversed for easy wiring of the six in series. Attach the metal switch parts to the balsa mounting strip, as shown, wrapping thread and gluing for strength. A bent pin serves as a rear clip and this is glued to the trailing edge of the wing at the center section.

Make two small holes in the motor mount, just outside of the square retaining strips. The ends of a short rubber band slip through these holes and are retained by straight pins. The motor fits under the rubber band. To save weight, cut off all but a short length of the lead wires. We are now ready to wire up the solar cells, switch, and motor.

Unbraid some stranded copper wire and use a single strand to hook up the power plant. Use a small electric soldering iron and do the job carefully. The solar-cell terminals nearest the motor go directly to the motor-lead wires. Notice that one is positive and one negative. Do not solder the connection at the motor yet, as we want to be sure the motor turns in the proper direction.

Using a short length of hook-up wire join the individual solar cells. Notice that negative joins positive all down the line. When you have wired together three cells on one side, run a wire to one side of the switch. Then join the other side of the switch to the terminal of the rear cell on the remaining bank of cells. Be sure to continue joining negative to positive.

Now we can check for correct polarity at the motor connection. Attach a rubber band from the switch to the rear hook. With the wing in sunlight or under a bright lamp, touch the two leads to the motor terminals. If the propeller blows air toward the rear of the wing the polarity is correct. If it blows forward, rotate the motor 180 degrees and then solder the final connections.

The wing may now be covered with paper. Use tissue except for the space around the motor, which should be covered with a piece of typewriter paper carefully trimmed to fit. When the wing is covered, it may be sprayed with water to tighten the paper. As soon as the water dries the entire wing should be pinned down to a flat surface with the wingtips blocked up the proper amount to prevent warping. Don't forget about the switch protruding from the bottom of the center section—pin the wing down to a board notched for the switch.

For proper lift the wing must set at the proper "angle of attack." Glue a triangular piece of balsa wood to the boom at the point at which the wing attaches. When the glue is dry, assemble wing and tail, using rubber bands, as shown, to attach booms. Support the underside of the wing center section with your fingers to check the balance. The plane should be level, fore and aft, when supported at the main spar. This indicates that the center of gravity coincides with the center of lift. If the nose is low, move the wing forward and check again. If the tail is low, move the wing aft and recheck.

Test glide the solar plane in tall grass if possible so that it won't be damaged in case of a dive or stall. If the plane dives, move the wing farther forward. If it stalls, move the wing back. Increasing the angle of attack of the wing also has the effect of raising the nose in flight, and decreasing the angle lowers the nose. You may want to experiment by changing the triangular strip of wood under the wing.

When you are satisfied with the glide of the plane it is time for the crucial test flight—under power. However, there is a safety measure we must take first. A rubber-band-powered model airplane has a limited motor run. So does a gasoline-powered type; in fact any conventional fuel will last for a limited time. But we have powered our solar plane with a most *unconventional* fuel! Without some means of switching off the electric motor we would have no way of controlling our solar plane and it would fly until the sun went down. For this reason we need a flight timer of some sort.

The mechanical flight timers used on gasoline-powered model airplanes weigh as much as or more than our complete plane, so we need a lighter method. We'll use the much lighter fuse-type timer that contest modelers have used for years to bring their craft down after a certain duration in the air. With chemically treated fuse material available at model-airplane shops we have a simple and lightweight timer for our solar plane.

Start the motor by stretching a rubber band between the switch and rear clip. Cut a short length of fuse and insert it in the rubber band. Light it carefully—you don't want to set fire to the airplane! In fact it may be

The author and the finished solar plane.

wise to light the fuse before inserting it in the rubber band. When it burns to the rubber band it breaks the band and stops the motor. You can thus regulate the duration of the motor run by using fuses of varying lengths. A small square of aluminum foil glued to the underside of the wing will protect it from any danger of being ignited by the hot fuse. Another wise precaution is not to fly your plane over an area where the dropped fuse might start a fire.

When all systems are go, launch your plane gently into the wind. This is best done by holding the booms with both hands just aft of the wing. If you have done your job well and the sun is doing its part your solar plane is on its way. There may be further refinements of balance required and you may want to warp a rudder slightly to adjust the circling of the craft. Remember, the flatter the turn, the more power will be received by the solar cells.

Another way to increase the power is to mount "solar chargers," as shown on page 122. As with our other projects these should face out 30 degrees from perpendicular to the solar cells. Make the solar-chargers from aluminized Mylar on paper backing, or use 1/32nd inch sheet balsa with aluminum foil cemented to it. The center reflectors are joined together at the top. Outer reflectors are braced with thin strips of bamboo or balsa wood. Trim the edges that attach to the wing so that they fit the curve of the ribs.

Measured with a volt-ammeter, the solar chargers

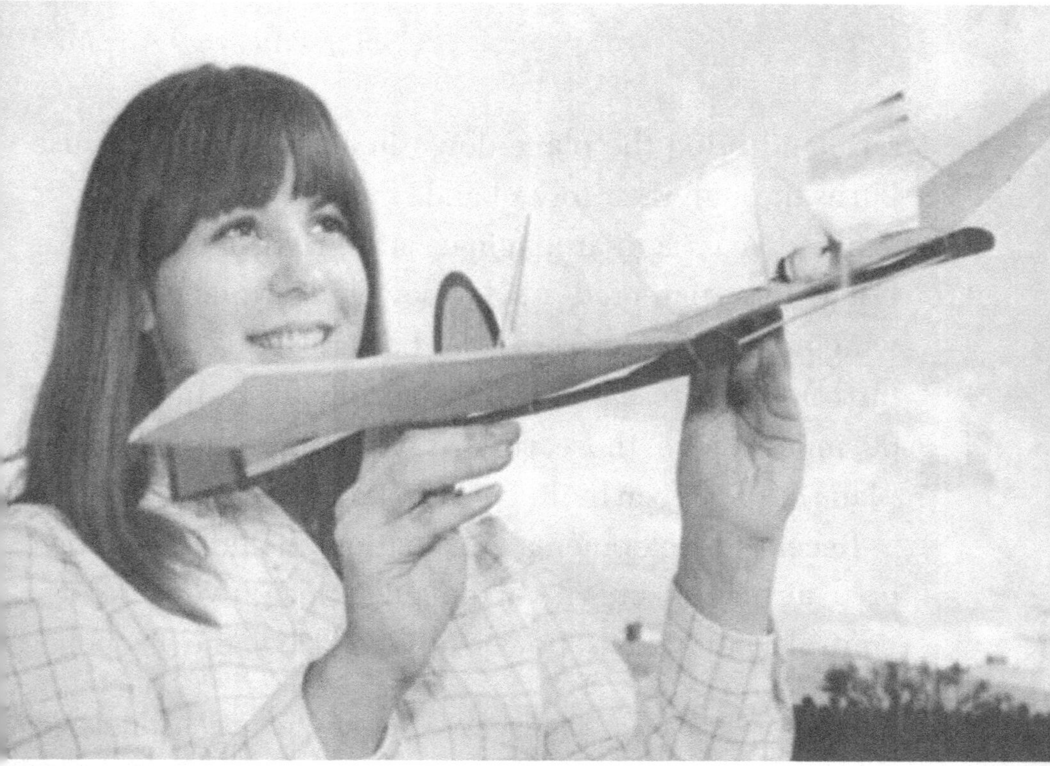

Solar plane fitted with aluminum foil "solar-chargers."

added more than 25 percent to the power of the solar plane. Of course, as the sun drops lower the reflectors may do more harm than good by shadowing the solar cells. Maximum power will be available when the sun is directly overhead in summertime. "High noon" is a good description of solar-plane flying. The higher the sun, the higher your craft will fly. After the motor stops, there is still another kind of sun power acting on your plane when conditions are right. To guard against rising thermal currents carrying your plane away you may want to operate a "dethermalizer" together with the motor switch. This is done by hinging the front edge of the stabilizer of the craft so that the tail may raise and bring the plane down in a stall when the fuse burns through the rubber band.

Some years ago an engineer at Wright-Patterson Air Force Base patented a solar-powered airplane. With a wing made up entirely of solar batteries, this unmanned, instrumented research craft would fly at altitudes up to 60 miles above the earth! As yet, this full-scale solar plane has not been built, but who knows?

Recent developments in aviation include manpowered airplanes, in which the pilot not only works the controls but pedals powerfully to drive the propeller

Futuristic unmanned solar-powered plane for high-altitude research.
SOLAR ENERGY SOCIETY

too. It has been found that this "one-manpower" engine can lift a plane, so the dreamers of old weren't really on the wrong track after all. Let's extend this finding a bit. One manpower is only a fraction of one horsepower, and thus one horsepower should fly a lightweight plane nicely. How much solar energy does it take to produce one horsepower?

Each cell of your solar plane develops a maximum of about one-third of a watt. Three of them produce one watt; and 3,000 cells would produce one kilowatt. A kilowatt is about 1-1/3 horsepower, thus 3,000 solar cells would be ample power for a lightweight plane. And these 3,000 cells have a total area of less than 200 square feet, about that of the wing on present man-powered planes. Furthermore, they would weigh only a few pounds. The only major problem of a solar-powered *man-carrying* airplane would seem to be its cost—something like 45,000 dollars worth of solar cells. With such a price tag the project might well be called shooting for the sun.

In the meantime you can enjoy solar flight for only a few dollars. It might be good insurance to put your name and address on your craft and offer a reward for its return—just in case it gets away from you some sunny day!

9

More Solar Projects

THIS BOOK SHOULD be only a starting point for anyone interested in solar energy and its applications. The field is still new, with lots of room for enterprise and imagination. Why not take over from these last pages and develop solar projects of your own?

If we can fly a model airplane with sunshine, there are obviously many other projects we can power with solar energy. Among them are toy cars ("sunmobile" would be a good word), boats, window displays, and so on. You can track such things down in back issues of popular mechanics and electronics publications. Or you can dream up some of your own. How about a solar-powered dirigible, for instance, using a gas-filled balloon and a solar cell or two to drive a motor and propeller? It would be a good idea to tether the dirigible unless you want to set some long-distance records.

In the model-car field you might copy the Baker electric car that International Rectifier Corporation

Solar panel atop old electric car furnishes power to storage batteries. Silent, pollution-free operation impresses Central Park policeman.
INTERNATIONAL RECTIFIER CORP.

once operated with a huge solar panel mounted on the roof. Or you could build a simpler, gearless, prop-driven racer. Attached to a central pole or mounted on a track, such cars could be raced in competition.

The light from electric bulbs will power solar motors too—not as well as the bright sun, of course, but enough for some applications. A word of caution here would be not to place a hot bulb too close to solar cells, since high temperatures can damage them.

Just as we could operate solar radios on stored solar energy, we can run solar motors. The answer lies in the rechargeable battery. A circuit drawing is presented as an idea of what can be done to provide power around the clock if needed. Actual solar cells, secondary batteries, and motors used will depend on the particular

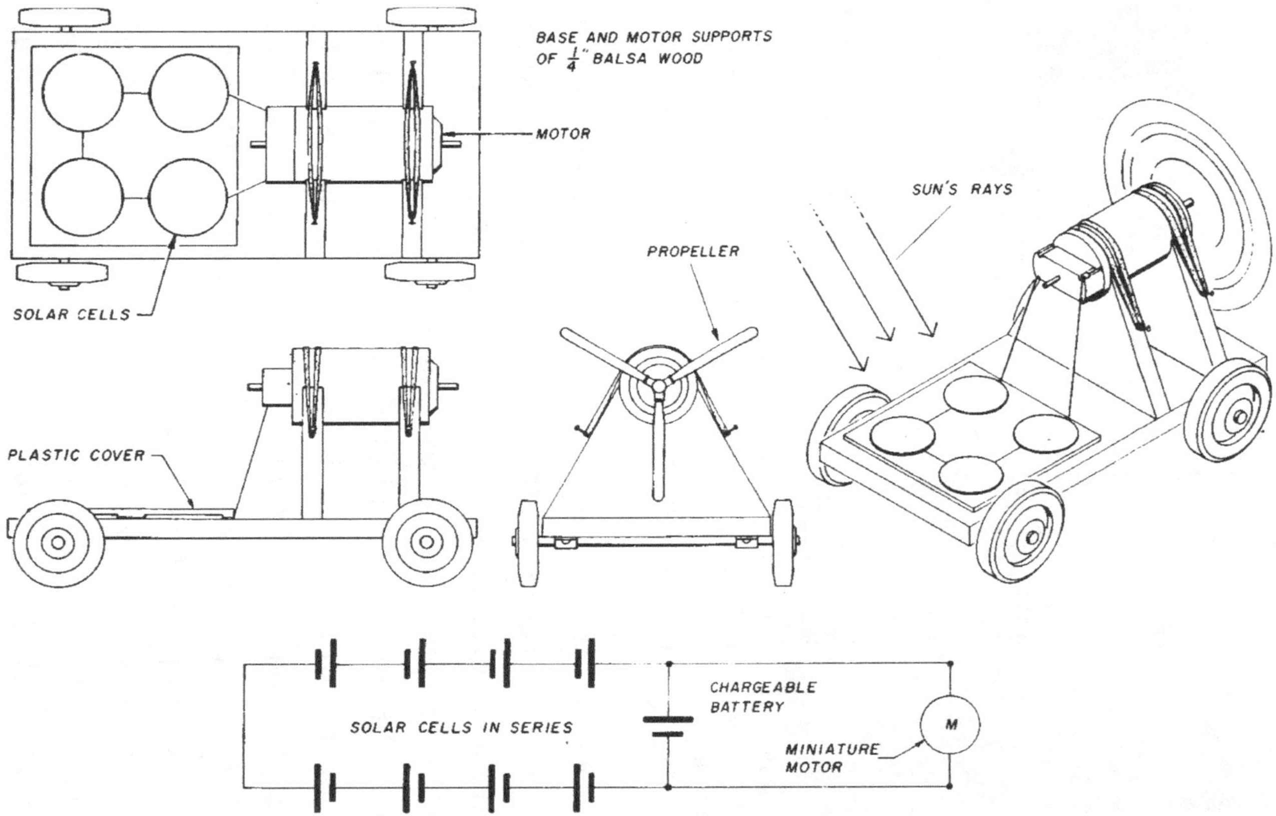

CIRCUIT FOR ROUND-THE-CLOCK OPERATION

SOLAR MOTOR

application you have in mind. The basic principle is the same: solar cells charge a chemical battery during the hours of sunshine to provide a reserve supply that can be used as needed.

There is probably a particular phase of solar work that appeals to you more than the others. If you are a radio or electronics fan, you'll spend more time on the chapter on the solar radios; then push on to a sun-powered transmitter or maybe a receiver with a loud-speaker instead of the simpler earphone used in the sets in this book. You can also build a variety of sun-operated relays or even a communication system that utilizes beams of light instead of radio waves.

If converting solar energy into mechanical power strikes your fancy, design and build a more powerful plant than the one presented here. Try a steam engine

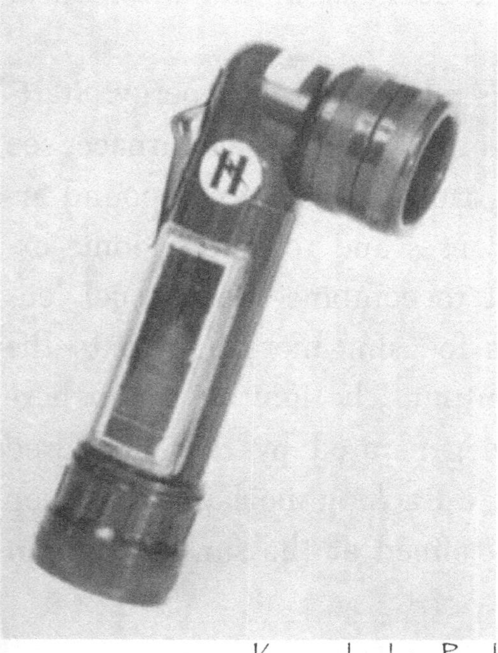

Solar-powered flashlight uses sunshine to recharge batteries inside the case.
HOFFMAN ELECTRONICS CORP.

Four solar cells drive this model aircraft carrier in its swimming-pool ocean.
HOFFMAN ELECTRONICS CORP.

with a solar-fired boiler. Solar turbines and hot-air engines have been made, and some of these have been presented in popular publications in the mechanical field.

The science of optics is related to solar energy. Starting with the Fresnel lens used in our solar furnace, see what other projects you can devise. Try compound arrangements with other lenses and reflectors. Some experimenters might want to combine the Fresnel lens with the solar cell, thus focusing more light onto the cell for greater power output. The limiting factor here will be the temperature generated by the condensed solar rays. You might try a tracking mechanism to keep a solar furnace properly aimed at the sun throughout the day.

The amateur chemist might be interested in the solar still, or perhaps a sterilizing device using sun heat. One researcher realized that there was a loss of efficiency in his solar still when drops of water collected on the glass. His solution to this problem was a special nitric acid treatment of the glass. This permitted moisture to condense in a film instead, resulting in greater transmission of solar heat. Another solar-energy application is the culture of algae: a food supply that some people suggest as a partial answer to the problems of food shortages.

There may be more applications for the simple emergency survival still described in the book. This prin-

Use of a sun-powered shaver is one way to beat five-o'clock shadow.
SOLAR ENERGY SOCIETY

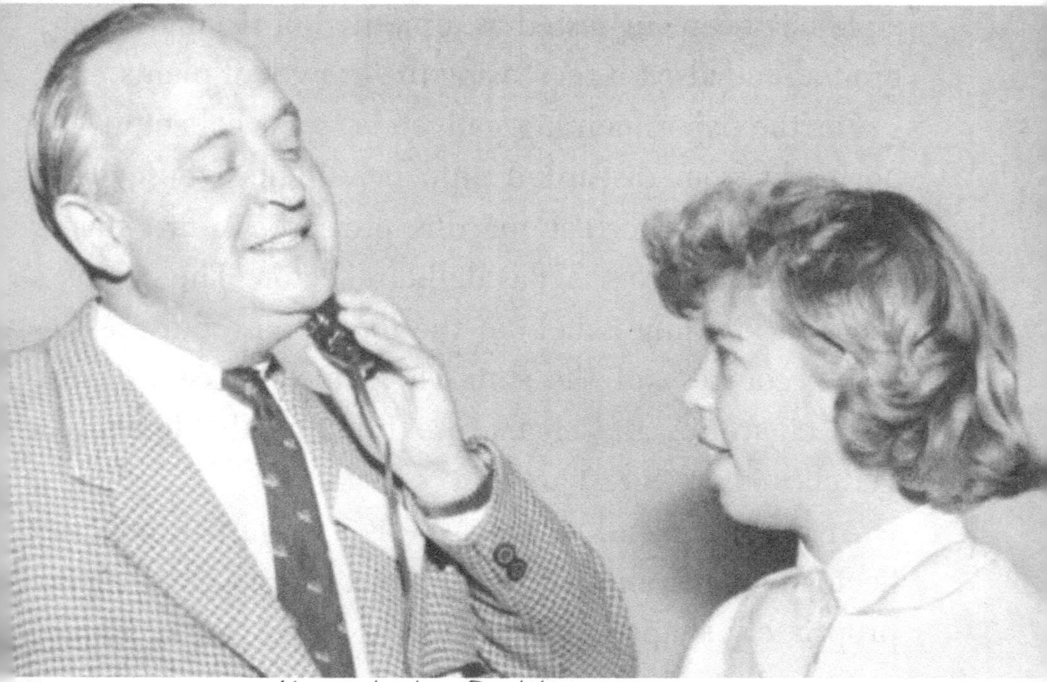

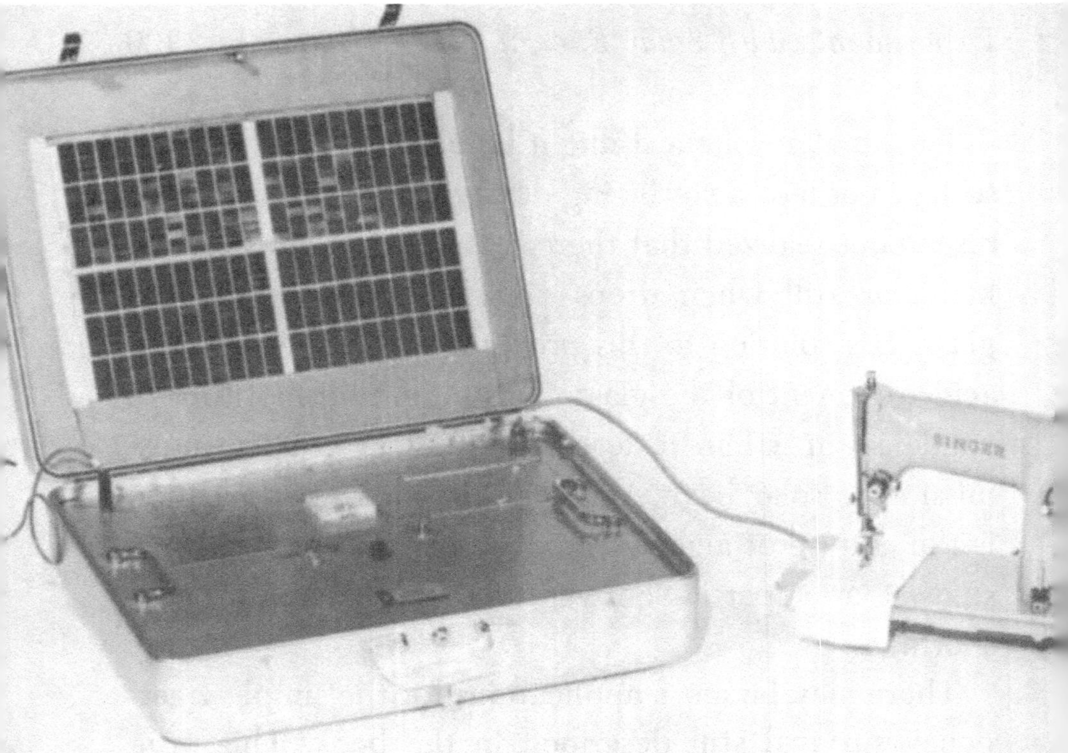

Since a stitch in time saves nine, why not take along this sun-powered portable sewing machine on that next trip?
SOLAR ENERGY SOCIETY

ciple has been suggested as a method of drying marshy ground and also for increasing the growth of plants.

For the experimenting gadgeteer, solar collectors can be made from discarded auto headlights whose parabolic shape is perfect for this purpose. A fire starter can be made from a flashlight reflector. The tip of a piece of kindling is held at the focal point and presto—a fire, courtesy of the sun. The same device will also light cigarettes and cigars, providing a safer light than matches in dry forest areas.

At the other end of the scale is a large parabolic reflector made from a surplus radar antenna or other appropriately shaped "dish." One of these would make a fine cooker or furnace, but keep in mind that the focal point may be very small and very hot. Once burned is twice careful, but that is a hard way to learn. Be careful too that you don't burn the bottom out of the cooking pot!

The solar device that probably goes back further in time than any other is the sundial. Although it is almost a curiosity now, it would be interesting and instructive to build one. Then be more modern and put together an electric clock powered by solar batteries.

Watch the newspapers and television for news of solar-heated homes or other such projects that may be built in your vicinity. If you are interested in architecture you already know how important the sun can be in helping to heat a home. Someday it will also be possible to cool buildings with sunshine.

Solar energy continues to have important applications in space. The sun is already providing the power for many of our lunar vehicles; perhaps it will soon operate power plants for the benefit of the first men on the moon. In the vacuum of the moon the sun's rays provide even more power than on earth, and the lunar day is two weeks long instead of only a few hours.

Perhaps the most poetic of all uses of sunlight would be that of "solar sailing" in space. In this book we present a model airplane driven by sunbeams and suggest, with tongue in cheek, a man-carrying craft powered by the sun. But some hardheaded space scientists have proposed to unfurl huge plastic sails and ride through

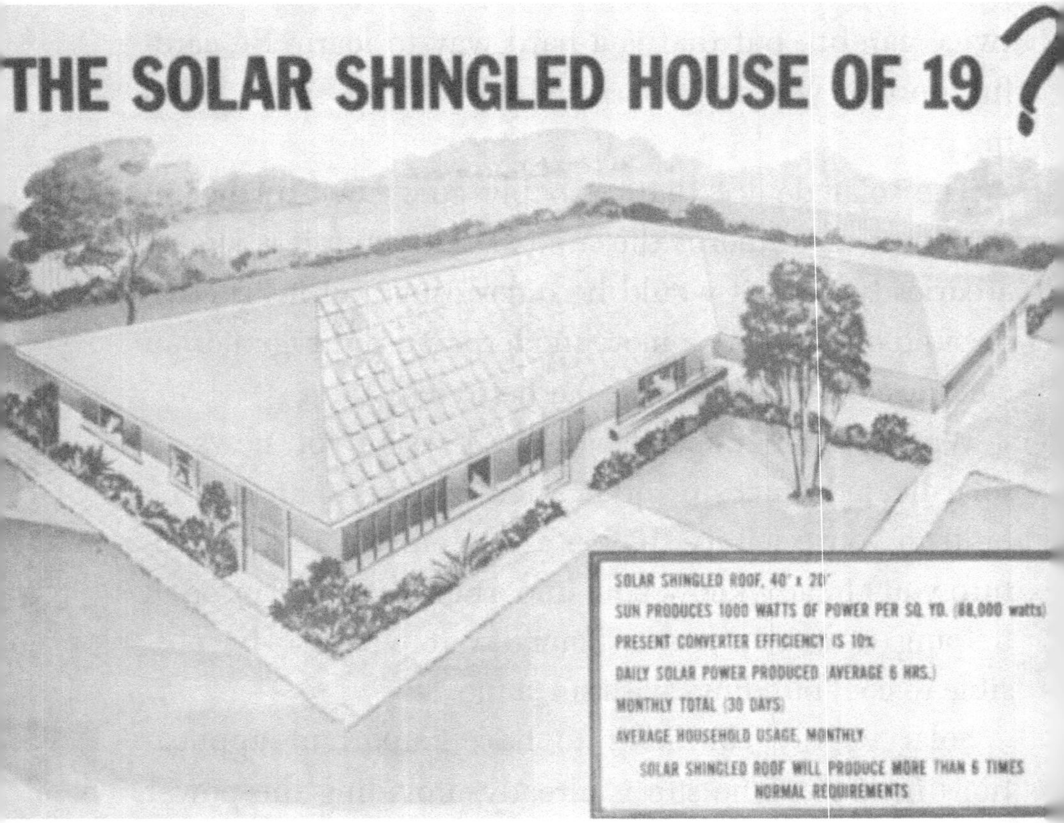

This fanciful solar home is not only heated, but is also provided with enough electricity for all household purposes. Solar "shingles" on the roof convert sunlight to energy.
HOFFMAN ELECTRONICS CORP.

Dr. T. C. Tsu studies a model of his solar sail.
WESTINGHOUSE

the endless reaches of space on the solar wind, a stream of tiny energetic particles. Whether such a dream will ever be reality is hard to say. But without it, there are still many wonders we will surely benefit from as science continues to make use of the bounty that is ours in the sun. Meantime, keep busy with the modest projects presented in this book.

The Sun and Solar Energy

Mass of sun	332,488 times mass of earth
Apparent diameter of sun	864,000 miles
Temperature of sun surface, approx.	10,000 degrees F.
Temperature of sun corona, approx.	20 million degrees F.
Temperature of sun center, approx.	30 million degrees F.
Solar mass consumed, per second	4 million tons of hydrogen (to helium)
Total energy release by sun	0.38 trillion, trillion kw (3.8×10^{23} kw)
Energy reaching outer atmosphere of earth	170 trillion kw (170×10^{12} kw)

Proportion of sun radiation reaching outer atmosphere of earth	less than one thousand millionth
Energy reaching earth surface, approx.	85 trillion kw (85×10^{12} kw)
Solar energy received in United States, per year, approx.	9000 trillion kw-hr or 1150 billion tons coal

(Solar energy reaching Lake Mead is five times greater than the power generated by the hydroelectric generators of Boulder Dam.)

Firms Engaged in Solar-Energy Applications

Bell Telephone Laboratories Murray Hill, New Jersey 07971	do-it-yourself solar cell kit
Clevite Corporation 540 East 105th Street Cleveland, Ohio 44108	cadmium-sulfide solar cells
Edmund Scientific Corporation Barrington, New Jersey	Fresnel lenses solar-furnace kits survival-still kits aluminum foil solar batteries electric motors

Education Materials & Equipment Co. Box 63 Bronxville, New York	solar-demonstration kit
Elgin Watch Company Elgine, Illinois	solar clock
Fabrite Metals Corporation 205 E. 42nd Street New York, N. Y. 10017	mirrors
Hoffman Electronics Corporation El Monte, California	solar cells solar-cell handbooks
International Rectifier Corporation El Segundo, California	solar cells solar-cell handbooks
Dr. George Lof Farmers Union Building Denver, Colorado	reflector cooker
Solar Energy Applications Laboratory Melpar, Inc. Falls Church, Virginia	various solar products
Micro-Mo Electronics Box 3952 Cleveland 20, Ohio	small electric motors

Howard W. Sams & Co., Ltd. 4300 West 62nd Street Indianapolis 6, Indiana	*Solar Cell and Photocell Experimenters Guide*
Solar Products Opa Locka, Florida	solar water heaters
Sundu Solar Heater Mr. Anthony J. Meagher 3319 Keys Lane Anaheim, California	solar heaters
Dr. Harry E. Thomason 6802 Walker Mill Road S.E. Washington, D. C. 20027	solar homes solar tepee
Yellott Solar Energy Laboratory 9051 North 7th Avenue Phoenix, Arizona	solar instruments research and testing
Zenith Sales Corporation 1900 North Austin Avenue Chicago, Ill. 60639	solar radios

Bibliography

Daniels, Farrington, *Direct Use of the Sun's Energy.* New Haven: Yale University Press, 1964.

Halacy, D. S., Jr., *The Coming Age of Solar Energy.* New York: Harper & Row, 1963.

———*Fabulous Fireball.* New York: The Macmillan Co., 1957.

Solar Energy. Tempe: Solar Energy Society, Arizona State University.

Sun at Work. Tempe: Solar Energy Society, Arizona State University.

Index

Folios in italics indicate illustrations